TINKERLAB

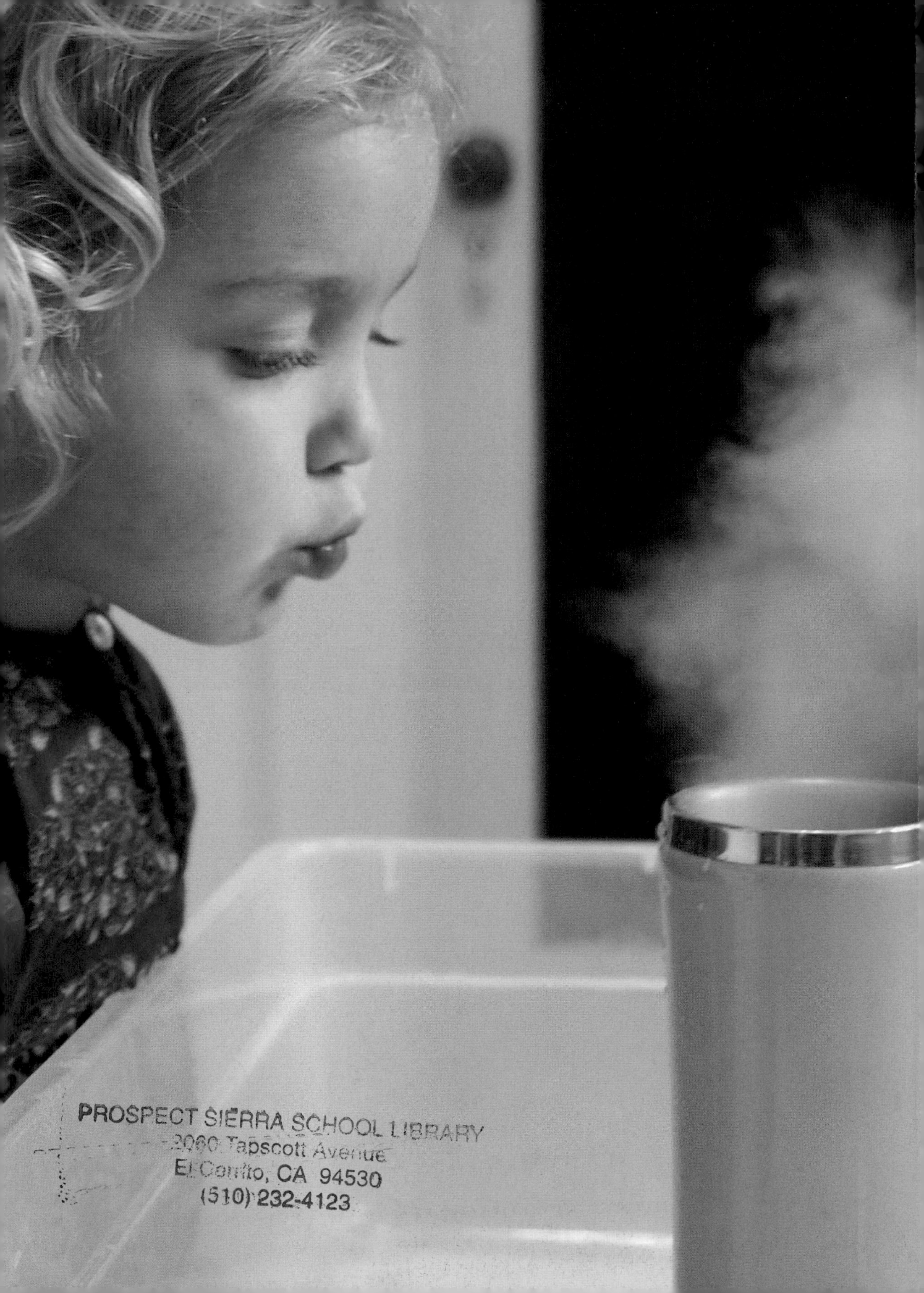

Tinkerlab

A Hands-On Guide for Little Inventors

RACHELLE DOORLEY

ROOST BOOKS

Boston & London

2014

Safety Note: The activities in this book are intended to be performed under adult supervision. Appropriate and reasonable caution is recommended when activities call for the use of materials such as sharp scissors, hot glue, or small items that could be choking hazards. Although this is a workbook, the recommendations in the activities in this book cannot replace common sense and sound judgment. Observe caution and safety at all times. The author and publisher disclaim liability for any damage, mishap, or injury that may occur from engaging in the activities in this book.

ROOST BOOKS
An imprint of Shambhala Publications, Inc.
Horticultural Hall
300 Massachusetts Avenue
Boston, Massachusetts 02115
roostbooks.com

Text and photographs © 2014 by Rachelle Doorley

"10 Lessons the Arts Teach" by Elliot Eisner was originally published in *The Arts and the Creation of Mind* (New Haven, Conn.: Yale University Press, 2002), 70–92. Reprinted with permission from the author.

9 8 7 6 5 4 3 2

Printed in the United States of America

♾ This edition is printed on acid-free paper that meets the American National Standards Institute Z39.48 Standard.
♻ Shambhala makes every effort to print on recycled paper.
For more information please visit www.shambhala.com.

Distributed in the United States by Penguin Random House LLC
and in Canada by Random House of Canada Ltd

Designed by Lora Zorian

Library of Congress Cataloging-in-Publication Data

Doorley, Rachelle.
Tinkerlab: a hands-on guide for little inventors/Rachelle Doorley.—First edition.
Pages   cm
Includes bibliographical references.
ISBN 978-1-61180-065-4 (pbk.: alk. paper)
1. Inventions. 2. Creative activities and seat work. 3. Playrooms. I. Title.
T47.D646 2014
600—dc23
2013027910

To my family; inspiring friends and
readers who enabled this journey;
and the creative and confident child
that's inside all of us.

Nothing is a mistake.
There's no win and no fail.
Only make.

—Corita Kent and John Cage

CONTENTS

5. Build 99

INTRODUCTION

All life is an experiment. The more
experiments you make the better.

—Ralph Waldo Emerson

It's early on a Saturday morning when I find myself in the kitchen with my children, watching my three-year-old whip up a batch of mystery pancakes. She's helped me make pancakes numerous times, and on this particular morning she decides that she's ready to select and mix ingredients on her own. She finds the mixing bowl, measuring spoons, flour, milk, baking powder, sugar, salt, and a few eggs. She drags out some cocoa, a bag of frozen blueberries, a banana, almonds, a mortar and pestle, and chocolate chips. Her focus is palpable as she proudly takes charge of the kitchen. Every now and then we exchange ideas about what should go into the batter, but for the most part, this is her show.

We cook the hotcakes together, sharing thoughts about shapes and cooking times, while keeping some distance between ourselves and the fire. As the pancakes come off the griddle, we devour them with a light drizzle of maple syrup. Admittedly, they're a little bit too sweet for my taste, but the experience is delicious.

If this sounds like it could be a scene from your home or one that you aspire to, you will probably take comfort in the ideas presented in this book. Making and experimenting are fun! This pancake breakfast represents hundreds of similar moments that have occurred in my home (of course, not all of them have gone as smoothly) that fall under the umbrella of *tinkering*.

WHAT IS TINKERING?

Traditionally speaking, tinkerers are those who take machines apart in order to figure out how they work, and then use that knowledge to innovate and build their own machines. For example, a tinkerer might wonder what makes a clock tick, take it apart to learn more, and then use what he or she has learned to make improvements on the clock design. This tinkering mind-set is no doubt responsible for great innovations in the design of appliances, cars, computers, and more.

My Uncle Mike, a child in the 1950s, recalls spending hours tinkering. When we chatted about his memories, he vividly remembered one instance of wanting to listen to baseball games that were broadcast from Los Angeles, sixty miles west of his home. His radio couldn't catch a strong signal, so he did what any resourceful child of the time would have done: he took his radio apart to learn how the parts worked and then rigged an elaborate scaffold of wires up the west wall of his room (hypothesizing that this would bring in a better signal) so he could hear his beloved team's games. And it worked!

At its core, tinkering begins with this kind of problem solving and a curiosity about how something works. When children are encouraged to solve problems on their own, they learn a great deal through the questions and hands-on experiments that lead to a solution. Even preverbal children pose questions and identify problems—think of a baby who works hard to grasp an out-of-reach toy.

THE TINKERING PROCESS

Tinkering is all about process. Generally, after a problem and the challenge of finding a solution are presented to children, you'll see a lot of discussion, tests, experiments, play, and yes, tinkering. The children may raise more questions, refine their ideas, or find their experiments are total flops. *But the results are not*

as important as the process. The process of being curious about something, asking questions, and exploring various solutions are all part of the fun of learning.

When children's improvements lead to a successful result, we might think of this as innovation. Regardless of the outcome, however, a good creative experience is one that gives children the opportunity to solve problems and think independently. When a child exercises flexible thinking in a project without a predetermined outcome, then the experience is a success.

HOW TO USE THIS BOOK

This book is designed for parents, teachers, and caregivers of young children who want to raise creative, independent thinkers. It is full of simple projects that foster creative thinking through hands-on experiences and provides an easy-to-follow guide to small habits, conversation points, and other tools to inspire a journey toward raising creative children.

First, we'll talk about how you can set up a tinkering studio and establish routines that foster and encourage creativity. Then we'll dig into the hands-on projects you'll enjoy in this space. If you already have a studio, feel free to skip ahead, but you might pick up a few new ideas if you thumb through the first section.

All of the experiences in this book are interdisciplinary in nature, but I've organized the projects into four main sections to help you easily find what you're looking for: Design, Build, Concoct, and Discover. You'll no doubt see crossover between the sections, and I encourage you to mix ideas and media as you forge your own path.

Within each of the four sections, the projects build on each other and become progressively complex. If you have a one-year-old, you might begin with an early project and then move on to the others when your child gains a few more skills. If you have a five-year-old, you might enjoy working straight through the sections or hopping around to pick projects that interest your child.

Children come to understand how materials work from repeated interactions with them. If you introduce a child to playdough, it's not enough to say, "Check! We've done playdough. What's next?" In his book, *Educating Artist Vision*, Elliot W. Eisner, professor of art education at Stanford University, writes, "Complex abilities, such as the production of art forms or the mastery of material, require the use of complex skills, and complex skills take time to develop. By moving quickly from one project with one set of materials to another project with another set of materials, the rate at which students must learn is increased considerably, and because such rates are generally unrealistic, children do not gain the type of competencies that breed confidence in their own work." Children learn, grow, and gain confidence by doing the same things over and over, and they ultimately develop competencies that would not emerge with one-off projects.

Lessons about the properties of paint and activities such as mark-making, tower building, and cooking all build on the same principle: The more exposure children have to a material or experience, the more they will learn what they can do with it. Children (and adults, for that matter) gain expertise through repetition. If you find a favorite project here, don't shy away from doing it repeatedly and then building on it to develop your own experiments. Some of these experiences may become household favorites that you do almost weekly, while others will be one-time wonders. Each child and family is unique and what appeals to or works for one may not capture the imagination or skill set of another.

Your goal is not to complete every activity in this book but to find materials, activities, and processes that your child enjoys exploring in depth.

As you read through the book, you'll notice that I have invited parents who have been in your shoes to share their firsthand experiences as counterpoints to my own. I was also lucky enough to hoodwink a handful of experts in the field to help us unravel some of the mysteries of creativity in early childhood. My hope is that these professional insights will balance my personal anecdotes.

LET'S GET STARTED

The projects shared in this book are intended to inspire you to infuse your children's learning experience with play and creativity. While some of the projects are child-driven and others require more adult guidance, the thread connecting them is that they all have inquiry and experimentation at their core. As children hypothesize, test, explore, and discover, they learn how to follow ideas to fruition and that there are multiple ways to solve problems.

So unbuckle your seatbelt and get ready to explore and try new things alongside your child. What's life, after all, but an experiment?

Prepare

TIP

Buy Art
Supplies
on SALE
in the Fall.

one

CREATING YOUR TINKERLAB

We shape our buildings;
therefore they shape us.

—Winston Churchill

A tinkerlab is a welcoming space that celebrates the processes of experimentation, exploration, and critical thinking. Turn your home into an environment that supports creativity and creation and you'll encourage your child to become an independent, flexible thinker and innovator who tackles life's problems and opportunities with gusto. Follow me, and I'll show you how to turn any space into a tinkerspace.

FINDING SPACE

*I don't really have studios. I wander around people's attics,
out in fields, in cellars, anyplace I find that invites me.*

—Andrew Wyeth, American painter

Spaces have the power to welcome us or push us away, and it's no surprise that Maria Montessori, an early childhood education innovator, refers to space as the third teacher. An inviting space can entice children to play and explore for hours, whereas a hostile space can shut them down.

What qualifies as inviting is different for all of us, and we each have space constraints that determine the scope and feel of a tinkerlab. You may have the ideal spot in your garage, dining room, playroom, back porch, or child's bedroom, even on a kitchen counter—the idea is to find something that works for you.

Think about your home. Is it tiny or big? Do you have a lot of storage, or are you busting at the seams? Do you already have a designated art area, or are you trying to figure out where you can carve such a space from?

We live in a small 1920s cottage. It's quaint, and at its cleanest, it's sometimes adorable, but mostly it's full of space challenges that I never seem to be able to wrap my head around. My husband, Scott, is the creative director at Stanford University's d.school, and part of his job centers around designing educational spaces for functionality and learning. But even with his expertise in the house, we're forever rearranging in an effort to figure out how our space can work best for us. Before we had children, the two of us would happily create in the extra bedroom (quickly turned into a studio) or at the dining table. But once children entered our lives and that extra space evaporated with all the furniture and stuff that comes along with kids, we realized we had to make room for creativity. But how?

Clear the Clutter

Two things have worked for us. One is purging. I come from a long, loving line of pack rats and mess makers. I attribute my crafty (and collecting) side to my grandmother Rosamond, who had a knack for bedazzling everything that entered her front door. As a child, I didn't see a problem with collecting every single shiny thing I adored, but this mind-set presented me with a lot of problems when I became a parent and my children's belongings quickly outnumbered my own.

I took a few weeks to clear a lot of clutter—all those things I thought I had to keep but didn't love or wasn't actually using on a regular basis. When I got rid of an oversized chair that I had planned to reupholster one day, I suddenly had room for a small art table. When I cleared out two drawers full of pens (gasp!) that accumulated over time, I was thrilled to have room to store my children's art supplies in a spot they could access easily.

Getting rid of nonessential items goes a long way toward everyday happiness—especially when you live in a small space. If you're the kind of person who likes to collect things, you'll appreciate what a hard, long lesson this was for me to learn. Oh, and lest you think this is a manifesto for small space living, the same holds true for big spaces: if you own something you don't love or use often, it's taking up space that could go toward something you really care about.

Need-Finding

Once I could see my space free of clutter, the second thing that helped us was comparing the way we wanted to use our house with the way we actually used it. In the world of design, they call this "need-finding." Scott and I took stock of our values and paid attention to our habits and how we used the house. This exercise helped us notice two things. One, we've always had a designated space to create things, and this makes us happy. As much as we enjoy eating meals at the dining table or watching TV from the couch, the most valued use of space for our family is a creative one. With this in mind, we knew that setting up a maker space trumped the importance of a traditional living room or dining room. Two, we noticed that our young children enjoyed spending most of their time in the bright dining area just off our kitchen. The space was

dominated by our big dining table, with just enough room left over for a few baskets of toys. It was hardly a creative zone, but if we moved the dining table out, perhaps it could be.

The result of this thinking prompted us to move our dining table into the living area and then turn our dining space into a creativity zone for the kids. My girls now enjoy working near me while I take care of things in the kitchen. Making this simple change increased the amount of time they make/build/tinker while also increasing my home management productivity.

While it's not uncommon for my kids to draw at the dining table or build towers in their bedroom, having a dedicated creative space reflects our family's values and enables my children to put their ideas into action with very few barriers.

If you find yourself in a similar situation, I encourage you to ask yourself these questions:

- How do we want to use our home?
- What is our home missing?
- How are we actually using our home?
- What areas get the most traffic? Why?
- What could we move, purge, or shift to make our dream home a reality?

Whether you have a small space like we do or more room to spread out, children are naturally creative and don't need a lot of fancy bells and whistles to stoke their imaginations. If you do have an extra space, such as a laundry room, converted garage, or playroom, by all means embrace it and make it an inspiring place to create. And if you live in a small home, take comfort in knowing that a studio can be carved out of any space.

Stimulate the Senses

One more thing to consider when selecting a spot for your tinkerlab is the overall feel of the space. Take a moment to think of a space that makes you happy.

What is it about this area that you love? How does it smell? What do you hear? What can you feel? Close your eyes for a moment and try to conjure up a clear picture of how your senses engage with this environment.

Having a source for music, an open window that brings in the sounds of birds and airplanes, the smell of warm-baked bread from the nearby kitchen, or a soft rug for cozying up with a sketchbook all provide kids with touch points to their learning experiences and childhood memories. I notice that my mood—and my children's—greatly improves when the stereo is playing or when we take our projects outside on a warm day. If memory is "the diary that we all carry with us," as Oscar Wilde said in *The Importance of Being Earnest,* then we should consider all the ways in which we're building this diary of time. Creating a comfortable and inviting space will help your children feel safe and provide them with an environment where they thrive, learn, and create.

Of course, once you have a dedicated studio space, don't believe for one minute that this will become "the spot" your kids will choose to create in. Frustrating, I know. All that hard work and they want to paint in the bathroom! Children move around and get inspired to create in all sorts of places. Be open to moving things around when this happens; their interests also change as they grow, so you'll be forever reconsidering how the space functions.

TIPS FOR CLEARING CHILDREN'S CLUTTER

Jillian Maxim

Get the kids involved in the process of organizing—it teaches them skills they can use for a lifetime.

What is enough? Begin a conversation with your children about the concept of enough. What is enough? Define *enough* for your family and stick to it. Examples: how many marking pens, craft supplies, stuffed animals, and so on are enough. Donate "extras" to children who don't have enough.

Sort items. Go through your children's items and ask them about which possessions/activities are currently the most important to them. Define and prioritize space for these items first. Encourage your children to let go of some of the leftover items; if they aren't ready yet, move these items to a less accessible area.

Define spaces. Organize and adapt your children's space according to what is currently most important to them. Define the spaces so that each child knows exactly where things go. This makes cleanup a snap.

Have fun with it! Remember that each time you sit down with your children to go through their stuff, you are not only clearing clutter and making space, but you are also giving them a chance to declare who they are at the moment.

HOW TO SET UP YOUR SPACE

Once you've established the space for your tinkerlab, you'll need to put some thought toward setting it up in a way that helps you raise a creative and independent thinker who's motivated to take action on his own ideas. I'll share two suggestions to get you started: one on creating a space that encourages children to follow their ideas, and another to stimulate the exploration of new ideas.

The Self-Serve Space

The key to setting up a space that encourages creativity and experimentation is to make it child-friendly by establishing self-serve zones where your kids can access safe and engaging materials on their own. If children can get what they need when they need it, they're empowered to act on their ideas. In addition,

hunting and gathering materials is a big part of the creative process. Discovering the "just right" piece of paper or selecting a specific color of marker enables children to make meaningful choices that have a real impact on their work. When they are free to choose materials that interest them or spark their imagination, they become actively engaged in the learning process.

While your studio setup may not seem terribly important, the benefits of creating a self-serve space can be profound. Children learn to ask and test their own questions; they seek out materials that help them realize solutions to self-designed problems; you are treating them with respect based on the belief that they can accomplish what they set their minds to; and finding materials on their own saves you time and supports their independence.

Despite the stereotype of artsy people being disorganized, organization does breed creativity. The way you organize supplies in your tinkerlab can go a long way toward inspiring your children to create, and it can be accomplished in just three steps:

1. Place everyday materials such as markers, tape, and crayons in an easy-to-access spot such as a low shelf or cabinet that children can open easily.
2. Fill clear boxes, bowls, or baskets with building materials such as wood scraps, corks, pipe cleaners, or straws.
3. Keep the art table clear so that it invites children to sit down and invent.

The self-serve zone will change as children grow. When my youngest child began to pull herself up on furniture, it was time to move all of the art supplies my older child was using to higher shelves. For this brief period, my older daughter could either ask me for what she needed or pull a step stool over to the shelf and find it herself. The additional challenge of having children of different ages and developmental stages is worth mentioning. While I might trust my older daughter wholeheartedly with permanent markers, my youngest doesn't have the long-term memory to recognize that permanent markers are different than crayons. The self-serve space still works beautifully with children of different stages, but not all materials are created equal, so take a few minutes to remove any supplies that could pose a danger to smaller children.

The added benefit of the self-serve space is that it makes cleanup considerably easier. If you want your child to create things, but you can't keep up with the mess, the best strategy is a proactive one: with a self-serve work space, the messes will still happen, but they'll be easy to contain, and tidying up won't feel like an insurmountable task.

Dealing with Messes

Although messes don't generally bother me, I feel like I'd be failing you if I didn't mention this elephant in the room. Taking issue with cleanup after a tornado painting session is one of the reasons I've had parents tell me they don't like to do art projects with their children. As much as I'd love to say, "Embrace the mess" (and I have said that publicly), I recognize that not everyone is comfortable with this strategy. There are times when my girls want to paint—like when we're about to leave for school—that just aren't good for me. Here are some tips for making your space a little less messy.

Make Your Self-Serve Materials Work for You. I like to acknowledge that there are good times to make messes and better times to make messes. For this reason, all of my children's self-serve materials are those "yes" items that I'm comfortable with them using anytime. They're things that can be swept up, gathered, and packed away easily when we're done. As the girls get older and better at cleaning paintbrushes or printmaking brayers themselves, I'm sure that my "yes" materials will grow to include these supplies as well. But for now, the materials my children have access to are those things that I know will create a manageable mess.

Everything Has a Place. Filling baskets or clear bins with materials also makes for easy cleanup. If a child knows where things go, he's more likely to put them back when he's done. You could add labels to the bins to make this step even easier. Put all your messy supplies out of reach for when *you're* in the mood for mess making, but make lots of "neat" supplies accessible so your child can create when the mood strikes.

Place a Wastebasket in Your Creative Zone. If my girls have to actually stand up and walk ten feet to throw something away, guess what? It will never happen! When it finally occurred to me to place a wastebasket under their art table, this all changed. I may still have to remind them to clean up, but because it's so easy to brush scraps off the table, they're more willing to help me keep the space tidy. I have a friend who keeps a handheld vacuum in her creative space, and she swears by it.

Be Ready with a Towel. If you're doing a project that promises to be on the messy side (such as painting, printmaking, or kitchen concoctions), have a damp towel, a box of baby wipes, or a roll of paper towels handy to clean up spills or wipe sticky hands. One of my blog readers says she has a bowl of warm, soapy water nearby to clean up spills as they happen.

Wear the Right Clothes. Wear aprons or clothes designated for making messes.

Pull Out a Tarp. Cover your table with oilcloth or some other easy-to-wipe table cover. If your work area is over a rug, having a tarp that you can pull out and clean off will save you a lot of anxiety during messy creative sessions.

Take It Outdoors. If the weather is good, set your messiest projects up outdoors. Hose your child down when the fun is done. Alternatively, do all your messy paint projects in the bathtub with washable paint or with food coloring added to shaving cream. It all washes off your child—and down the drain—when the fun is done.

Involve Your Child. Children are almost never too young to learn how to clean up after themselves. To make this chore more fun, we have an arsenal of cleanup songs that we bust out when we tidy up. I inevitably end up doing more work than my kids, but the repetition of this process undoubtedly sinks in. In the spirit of Montessori teaching, we try to clean up one activity before moving on to the next, which helps motivate my children to be responsible for their messes. I think making them accountable for their spills helps add some control to the chaos. It's not foolproof, but it does seem like this extra layer of responsibility keeps them from making messes just for the pure thrill of it.

HOW DO YOU SET UP YOUR SELF-SERVE AREA?

I keep a "salad bar" style space in our art room. I use berry containers and jars to hold art supplies, trinkets, and recycled goods. My children can add to and take from the supplies as often as they please. My favorite art pieces that they have made come from this style of creating. They always make such beautiful abstract art without any influence except what is in their own little imaginations.

—Melissa A.

We have "making shelves" that are always stocked with all sorts of things the children can help themselves to: a box of "junk" for upcycling, papers, pen and pencils, scissors, tape, glue, and drinking straws. I add in extra items as the seasons change or I come across a new material for them to try. They are free to use the materials whenever they like and to do their making anywhere—favorite places are on the floor right in front of the shelves or on the kitchen table. The benefit of having these open-access shelves is huge. The children know what is available and can get materials independently, so I see them using their imaginations daily, incorporating the art materials into their role play, making models for their toys to use. . . . I love to use recycled materials and inexpensive items, which means no one needs to feel things are "precious" and too special to have a try with. I see them developing fine motor and spatial skills quite naturally through their making, and it lets talents that often go unnoticed at school shine through.

—Cathy J.

We have a kid-sized table in our kitchen—the "art table"—that has free access to paper, scissors, tape, crayons, markers, stamps and ink, and a number of other crafty items. I try to keep most of its materials in a nearby closet that the kids have access to, but markers, paper, and some seasonal items are always out on the table. We have markers for each child in labeled containers, several containers of crayons, and a stack of plain white paper. We also have a large box underneath the table in which the kids put works they decide to keep. (I like delegating retention decisions to them!) I love the creations/inventions/gifts they produce. The table is an automatic first stop at our house, and I love it for the creativity it inspires and, quite frankly, for the time it gives me.

—Sarah H.

We have a kid-sized table in my son's playroom, and on top of it is an art caddy with loads of art/crafty things: scissors, glue, "sticky tape" (as he calls it), paint, crayons, stickers, pieces of cloth, yarn, buttons, paper, containers with tiny pieces of paper that he can shred for future projects (I change/add supplies as needed). He loves the fact that everything is so accessible, and I can see he gets a sense of independence and pride knowing I trust him with more "grown-up" supplies like buttons and sewing needles.

—Paola L.

For their experiments we have converted my children's toy kitchenette into a small laboratory in the kitchen. The kids love it. The idea is to suggest using materials and document and see how they work. It consists of an area of observation: magnifiers, microscopes (currently

under my supervision), anaglyph 3D glasses, colored lenses, and glasses
. . . and our laboratory has an area of exploration/experimentation:
measuring cups, spoons, beakers, droppers, tweezers, scissors . . . and
materials to experience buoyancy, effervescence, solutions, miscible liquids, hydration. . . . They love to wear gloves and safety glasses
while using their notebooks and pens to document their thoughts.

—Àngela A.

I am totally afraid of mess! Our house is tiny and any additional clutter seems to get under my skin. My favorite place for art projects is
outside in our driveway. I bring out the kids' table and chairs and set
them up, then bring as much paint, glitter, glue, and paper as the kids
want. They're happy, and I'm happy.

—Aleksandra D.

Creative Invitations

Creating a self-serve space is a wonderful way to make tinkering an always-available option for your children. You can build on this idea by coming up with creative ways to offer them new materials and creative opportunities. One way to do this is with creative invitations. This is an idea that is often used in preschools.

The children who walk into classrooms at Stanford University's Bing Nursery School are welcomed by a handful of thoughtful invitations, or "provocations," that invite them to create and play: a table with small mountains of clay, a water table filled with rubber fish and fishing nets, or a sandbox "excavation site" filled with tall shovels and a running hose to make rivers. The children are intrigued by these invitations, and it's only a matter of minutes before they find themselves engaged in a new world of possibilities.

We can learn from years of preschool teaching wisdom and re-create the magic of creative invitations in our own homes.

Quite simply, a creative invitation is a combination of materials and context that intrigue children with a suggestion of play. In the preceding water table example, the addition of rubber fish and fishing nets to the water invites the children to scoop the toys from the water, pretend they're fishermen, or playact a story about fish and mermaids. While the materials may imply something specific, there's no expectation placed on the children, and the outcome is completely open-ended.

Deb Curtis and Margie Carter offer some guidance on selecting materials for creative invitations in their book *Designs for Living and Learning: Transforming Early Childhood Environments:* "The presentation of materials makes a difference in how children respond to them. Make sure the arrangement is orderly and attractive and that it suggests possibilities for use. Baskets, trays, tubs, mirrors, or other surfaces define the area and help children focus their attention on what is available. Avoid the cluttering effect of combining different-looking implements and utensils together in one arrangement. Offer sets of things that match and complement each other so the children have a clearer view of what is there and how it may be used."

Creative invitations don't have to be complicated; in fact, they're usually better if they're not. The invitation can be as simple as moving the painting easel to a different room or placing a bowl of plastic eggs and spoons next to a tub of water. In essence, the idea is to encourage the growth of new ideas. Sometimes the invitation will require an explanation and often it won't. When you set up a provocation, you may have some expectation of how the project will develop, but the beauty is that it can grow in any number of directions.

Here are some tips on how to present a creative invitation.

Make the Invitation Beautiful. Clear away anything that's irrelevant to the invitation. Organize it with nice wooden bowls and clear containers, or place it on a clean tablecloth.

Collect Multiples of One Material. Having a variety of one item is always interesting and encourages pattern making and experimentation. Some examples include a bowl full of pebbles and an empty muffin tray, a tray of circular stickers and an appliance box, and a basket of mushrooms and a kid-friendly chopping knife.

Put Familiar Materials Together in New Ways to Suggest Alternate Uses. If your child usually draws with crayons on white copy paper, cover the underside of a table with paper, move the bowl of crayons to the floor, place a few pillows under the table, and invite your child to draw upside down.

Offer a Limited Selection of Materials That Complement One Another. Too many choices can overwhelm a child, so keep the options to a bare minimum.

Follow Your Child's Interests and Use Them as a Starting Point. If your child is fascinated by rockets, maybe she'll enjoy the challenge of working with an upcycled water bottle with windows cut out of it, small squares of colorful paper, tape, and a ball of string.

Gather Materials Ahead of Time. To set up an invitation, take a few moments to gather your materials ahead of time. If you enjoy greeting your child with early-morning inspiration, lay everything out the night before. If you're planning this for later in the day, think about the invitation ahead of time and then take a few minutes to clear some space and set out the materials. Open-ended invitations should not take a lot of time to set up, but they may require a bit of foresight.

Creative invitations have been one of my best discoveries for fostering early childhood creativity as a parent. They're not prescriptive, and my children can interpret them as they please. More than anything, children generally enjoy the challenge of combining materials in new ways, and invitations like these can help parents avoid the stress of coming up with a picture-perfect craft.

WHAT NURSERY SCHOOLS CAN TEACH US ABOUT CREATIVE INVITATIONS

An Interview with Nancy Howe, Head Teacher, Stanford's Bing Nursery School

Q: When children walk into your classroom, they're greeted with a handful of inviting activities. Can you tell me about these creative invitations and how they support a child's development?

A: I love creative invitations! We call them "setups," or if we are being a bit more intentional in terms of possible outcome, we might create a "provocation." I think some of us who spent time studying the Reggio Emilia approach in Italy were inspired by that concept. A creative invitation or setup from a teacher's perspective would involve making the activity truly inviting to children. For example, with playdough, making sure that it is the right texture for pliability—not sticky or dry. We start simply so that children can explore the affordances [qualities] of the material. Minimize the distractions. Just hands at first. Maybe even no color. Tools and accessories can come later. Poking, patting, and rolling can be modeled by an adult before, during, or after children have had a chance for self-exploration. I can see the advantages of all.

Teachers can help children learn when they scaffold knowledge, or break down learning objectives into developmentally appropriate chunks. Repetition of an activity also fosters opportunities for further, deeper exploration and innovation. "Boring" can be a good thing!

I think parents encourage immediate gratification by going for all the bells and whistles right away; then children aren't satisfied with the lack of them, with the basics. When things are stripped down to their basic elements, it encourages children to add their imaginative input. I see it as a sort of creative survival. A provocation with playdough might be as simple as putting a big ball of playdough out and letting children help themselves—or use the whole thing. Adding a novel tool, another texture, color, several colors to mix. Perhaps letting children mix their own playdough without a recipe.

As teachers, we are also involved in the social dynamics of an activity and our setups reflect this. We might initially set out a hand-sized ball of playdough for each child, as well as a knife and a roller. It's always interesting to see how children negotiate tools and materials in an open-ended and ongoing activity. They often need support around social interactions rather than technique.

Q: What elements need to be in place for a provocation to be successful?
A: Thoughtfulness, intention, and a willingness to be flexible and comfortable with the unexpected. Knowing how to respond to children's work, especially the process and techniques they use rather than just the finished product, gives children confidence in their ability to trust their creativity and imaginations.

Q: What are your favorite materials for open-ended exploration?
A: I love all the "basic" materials we use at Bing (paint, clay, sand, water, and blocks). They all tap into different areas of children's creativity. Recycled or found materials should also be included, since they provide so many opportunities for creative interpretation and are so readily available. They really encourage children to look at everyday objects in a new way.

Q: What are your top three tips for parents who want to set up a successful play invitation for their two- to five-year-olds?
A: 1. Allow for lots of time for exploration of materials.
   2. Engage with your children by providing nonjudgmental comments about what they are doing. For example, "You're poking lots of holes."
   3. Ask good questions to expand their play. For example, "I wonder what would happen if . . ."

TOOLS FOR TINKERING

If you have the right supplies and tools ready for your small inventors, their ideas can be realized with few obstacles. I started our supply cabinet when my eldest was almost two, and after a lot of trial and error, here's what I've learned:

- The best materials are the simplest.
- Group like materials together.
- Making your own supplies can be worth the time invested.
- Start off with just a few things and add as you go.

To make it easier for you, I've broken down the materials you'll need into two main categories: design supplies and discovery supplies. Design supplies are used to invent, build, tinker, and make. Examples are tape, paper, wood, and a glue gun. Two project sections of this book, "Design" and "Build," call on these supplies, and the design and tool kits described in this chapter are guidelines to help you get started. Discovery supplies are used to learn about the natural world, make scientific discoveries, and explore the senses. The "Concoct" and "Discover" project sections of the book will call for discovery supplies such as seashells, measuring cups, and a light table. Of course you'll notice a lot of overlap in materials from project to project, but having a clear sense of what you may need can help you stock your supply cabinet.

IDEA JOURNAL

The backbone of any kind of tinkering is an idea journal. It's not just useful for drawing; it's also a great place to collect mementos, write ideas, and explore scientific discoveries. These ongoing memory keepers provide my husband and me with a peek into our children's growth, and when my girls pick up their journals, they often begin by flipping through the pages to see what they've done and as a source for new ideas. Large, unlined, heavyweight paper notebooks are ideal for absorbing wet media such as paint, give children enough space to write or draw, and provide them with room to save ephemera such as maps or birthday cards. I highly recommend giving your child a notebook in which he can note his ideas, experiments, discoveries, and thoughts.

DESIGN SUPPLIES

When graduate students walk into the prototyping area of Stanford University's Hasso Plattner Institute of Design (aka the d.school), they're greeted by a large carousel full of well-organized and inspiring design supplies such as string, vinyl letters, colorful tape, decorative paper, pipe cleaners, and foam. When the students get a challenge to design something, they'll often use this cabinet of curiosities for inspiration. This carousel may seem like a mishmash of ideas, but it's thoughtfully stocked with five types of materials that d.school students use for quickly modeling representations of ideas that can be worked out and tested: utensils, pliable materials, connectors, structural items, and treasures. With a little bit of effort, you can easily re-create this d.school magic right in your own home.

If you consider some of the most basic children's craft supplies, they fall into the same five categories: crayons (utensils), clay (pliable materials), tape (connectors), pipe cleaners (structural items), and crafty bits and pieces (treasures). To stock your pantry, pick a few materials from each category and display them in an inviting way where children can reach them on their own. Here are some ideas:

- Utensils: brushes, pencils, crayons, markers, colored pencils
- Pliable materials: paper, rubber bands, playdough, aluminum foil
- Connectors: tape, staples, glue
- Structural items: toothpicks, popsicle sticks, cardboard tubes
- Treasures: recyclables, ribbon, balloons, sequins, pom-poms

HOW TO CHOOSE ART MATERIALS

With so many materials to choose from, I'm often asked for recommendations on which kind of paper is best for watercolor painting or what marker brand is my favorite. The truth is I'm not all that picky. When selecting art supplies, I follow these three basic guidelines:

1. Buy the best quality you can afford.
2. Always go with nontoxic. Look for products that the Art & Creative Materials Institute (ACMI) certifies with its AP (approved product) or CL (certified label) seals.
3. Washable anything will not disappoint.

The Design Kit

This kit is limited to the basics that encourage independent art making, and it's a great place to start if you want to keep it simple.

Basics

- A stack of paper or a sketchbook
- Markers, colored pencils, and/or crayons
- Scissors
- Playdough
- Liquid watercolors and a brush
- Stickers
- Glue
- Tape
- String
- Treasures

Extra, Fun Stuff

- Oil pastels
- Stencils
- Rubber stamps
- Brayers
- Printmaking ink
- Permanent markers
- Box of found papers
- Box of recyclables
- Tempera paint

The Tool Kit

Any tinkerer worth his or her salt has to have a tool kit. But what goes in it? While you can find child-ready tool kits online and in some of the bigger hardware stores, if you'd like to assemble your own, this list of essential tools can help. To keep your kit kid-friendly, make sure you get the smallest size tools you can find. I'll share some guidelines, but available sizes will vary.

Basics

- 6-ounce claw hammer
- Measuring tape
- 6" long-nose pliers
- 6" flat-head screwdriver
- 6" Phillips-head screwdriver
- Pack of miniscrewdrivers
- Low-heat glue gun
- Scissors

- Wood glue
- Two 2" clamps
- Pencil

- Ruler
- Safety glasses
- Carrying case

Extra, Fun Stuff
- Flashlight
- 6" adjustable wrench
- 9" magnetic torpedo level
- Child-sized leather gloves
- Manual hand drill (looks like an eggbeater)
- Workbench vise

DISCOVERY SUPPLIES

When children hike in the woods, skip stones at the shore, or fish in a creek, they are given the best shot at building an understanding of how their lives are interconnected with the natural world around them. These personal interactions help them ask meaningful questions about nature and give parents the opportunity to share and point things out in a way that's more or less unattainable through books alone. To support and build on such interactions at home, it

helps to have an area or a shelf that's dedicated to nature-based discovery. The materials in this kit will change frequently depending on the season, your location, and your child's interests. For example, after a trip to the beach, we might bring home driftwood and broken pieces of coral to investigate or display. While some of the materials on the following list can be purchased, the primary focus here is to explore natural materials such as rocks, pinecones, and leaves. If you want to purchase just a few things, I'd start with a magnifying glass, microscope (for looking at pond water or leaves), a camera, a few empty containers in which to collect your treasures, and a notebook. Here's a more complete list, if you're looking for a few additional ideas:

Investigations
- Magnifying glass
- Tweezers
- Child-friendly microscope and slides (premade or blank; we have both)
- Bug jar
- Binoculars
- Clipboard, paper, and pencil

Concoctions and Measurement
- Funnel
- Measuring cups
- Clear jars
- Turkey baster
- Eyedropper
- Tumbled glass pieces
- Test tubes
- Lab coat (just for fun!)
- Safety goggles
- Trays to capture overflow
- Egg timer

Nature
- Book about animals or plants
- Seashells
- Rocks
- Crystals
- Acorns
- Leaves
- Pinecones
- Snakeskin
- Piece from a beehive
- Flowers
- Twigs
- Moss

Light
- Flashlight
- Light table
- Overhead projector
- Prisms
- Mirrors

Discovery Pack
(for on-the-road observations)
- Magnifying glass
- Bug jar
- Binoculars
- Sketchbook
- Colored pencils

WHAT ARE YOUR CHILD'S FAVORITE MATERIALS AT THE MOMENT?

My almost-two-year-old loves stickers, stamps, colored masking tape, Do-A-Dot art, and chunky pencils. He also loves wooden and Mega Bloks.

—Sara Y.

My two children are almost five. My son's favorite materials for creative play are Lego bricks. My daughter enjoys those as well, but she's been drawing more lately, which means she has been reaching for markers, crayons (with sparkles!), and that great 80-pound paper from Discount School Supply.

—Chelsea D.

My daughter is almost six, and her favorite is easily Stockmar modeling beeswax.

—Jen B.

My girls are four years, two years, and ten months old. The big girls are always smitten by paint of any variety and love to paint both pictures and junk materials. The four-year-old loves drawing now that she has discovered the magic, and my two-year-old adores snipping and sticking everything and anything (thankfully, not her hair or clothes yet!).

—Anna R.

My five-year-old loves things like FIMO clay, junk modeling, and drawing. My three-year-old loves painting and playdough.

—Maggy W.

My one-year-old loves "dawning," which means getting out the Crayola markers and letting her loose on some paper. The three-year-old is into collages, so glue and our bits-and-bobs box, and he's happy.

—Cerys P.

I have two six-year-olds and a three-year-old. All three love Mega Bloks and Legos, paint, and playdough. Scissors, markers, tape, and staples (with paper) are the favorites for the six-year-olds right now. For the three-year-old, it would be markers and stickers.

—Sarah H.

My two-year-old loves playdough, which he uses with modeling tools and lollipop sticks, matchsticks, large beads, plastic jewels, feathers, and googly eyes.

—Catherine B.

My two children are five years old, and at this moment, my son's favorite materials are *anything* he can use to make concoctions combined with droppers, measuring glasses, magnifiers, and food coloring. He also loves watercolors and markers, which he uses to make tiny paper puppets. My daughter's favorites are washi tape and *anything* she can use to create a costume. She also loves all of my permanent markers and my superdelicate pen brushes (the ones I try to keep away from their little hands).

—Angela A.

TEN TINKERLAB
HABITS OF MIND

Habit 1 Make room for creativity

Habit 2 Encourage questions

Habit 3 Listen actively

Habit 4 Be curious

Habit 5 See mistakes as gifts

Habit 6 Embrace a good mess

Habit 7 Accept boredom as a tool
for self-discovery

Habit 8 Step back and enjoy the flow

Habit 9 Spend time outdoors

Habit 10 Think of everything
as an experiment

TEN TINKERLAB HABITS OF MIND

As I reflect on what makes our tinker-space tick, ten habits of mind emerge as the standards of my parenting curriculum. While they're not in my head all the time, each of them plays a role in how I organize our space, talk to our children, and encourage their independent voices to flourish. Of course, children learn a lot on their own, but parents, teachers, grandparents, and caregivers can support learning by layering knowledge and experiences with carefully selected information and encouragement. With these habits as a guide, we can be more mindful about how we set up provocations and introduce our children to new ideas.

1. MAKE ROOM FOR CREATIVITY

I think of my studio as a vegetable garden, where things follow their natural course. They grow, they ripen. You have to graft. You have to water.

—Joan Miró, Spanish painter

Human beings are creative by nature, and the fluency and variety of our creative abilities may be the most meaningful characteristic that separates us from other species. We engage in creative activities all day long—from the simple act of choosing an outfit to the more complicated task of cooking a meal.

The first way to make room for creativity is to make sure that you, the grown-up, pursue your own creative interests. Do you like to sing, dance, paint, or perform? Do you make time for this? It doesn't have to be a lot of time, but it's important to begin with "you." When children see that the adults in their lives are on a creative journey, they'll view this as a worthwhile way to spend time.

The second way to make room for creativity is to make materials and a designated space available to children. Access to materials is key, but even more important may be setting aside lots of free time to explore ideas. What good is a thoughtfully designed creative space if children don't have time to dig deep and follow their ideas? Long periods of unstructured time to explore the same materials over multiple days helps children understand the properties and potential of those materials. As Bing Nursery School teacher Parul Chandra says, "If you have the materials out for only one day, how will children really get to the bottom of it?"

2. ENCOURAGE QUESTIONS

The scientist is not a person who gives the right answers,
he's one who asks the right questions.

—Claude Lévi-Strauss, French anthropologist

Self-motivation goes a long way toward learning, and one of the most important things parents can do is create a safe space for children to pose and answer questions. While we may carry around more information than our children, we can be better resources if we encourage them to ask their own questions as we guide discovery.

When children ask questions it shows that they are truly curious about something and are engaged in the process of learning. It's important to support this curiosity and equally important to return the favor by asking a child questions that can lead to bigger ideas and deeper thinking.

Conversations that honor a child's interests and are rich with questions can support their creativity. For example, if a child says, "I don't like the way this brush works," you can ask her to explain her reasoning: "What is it about the brush that you don't like?" or "What do you think would make it work better?" Questions like these help children think through and communicate their ideas.

But not all questions are created equal. When you ask a child a closed-ended question, you risk shutting him down, or you may fail to understand his point of view. Closed-ended questions suggest a yes or no answer or a specific one-word response. For example, "Are you drawing a sun?" or "What color is a stop sign?"

However, ask a child open-ended questions, and she'll share her ideas, reflect on her thinking, and understand that you care about her ideas. Open-ended questions have no expected answer. Improving on the two previous questions, we could try, "Can you tell me about what you're drawing?" or "What kind of street signs have you noticed in our neighborhood?"

Here are some types of open-ended questions that can engage your child in a thoughtful way.

Knowledge Questions. These could start with "What do you know about . . .?" For example, "What do you know about trees?" When a child is curious about something, begin by asking what he already knows about that thing.

Process Observations. These might start with "I noticed that you are . . ." For example, "I notice that you mixed green and white paint together" or "I see that you're connecting the tubes with tape." When your child is engaged in the process of exploration, comment on the visible qualities of his actions without making any judgments or interpretations. Be as objective as possible about what you see. Your observations show your child that you notice his actions and give you the opportunity to introduce him to new vocabulary words and concepts.

Reflection Questions. These might begin with "Tell me more about . . ." or "How did you . . .?" For example, "Tell me more about this purple shape at the bottom of the paper." Or "How did you connect these pieces together?" Since you're trying to refrain from interpreting your child's work, at least in an outward way, these questions invite your child to clarify her goals.

3. LISTEN ACTIVELY

Creativity becomes more visible when adults try to be more attentive to the cognitive processes of children than to the results they achieve in various fields of doing and understanding.

—Loris Malaguzzi, educator and founder
of the Reggio Emilia learning approach

There's listening, and then there's *active* listening. Active listeners make a conscious effort to understand what a speaker is trying to communicate and often paraphrase what they hear back to the speaker. This is a powerful tool for connecting with children and can deepen the parent-child bond. Active listeners support the speaker with verbal feedback while honoring the speaker's point of view. This feedback shows that the listener is paying attention, and for parents, it can also be an opportunity to introduce new vocabulary words or grammar.

When a child makes an observation about the weather or a comment about her lunch, an active listener might paraphrase these comments back to her. For example, if she says, "Look, it's raining!" an active listener might say, "Isn't that amazing? It was sunny a moment ago, and now it's turning into a wet day." If she says, "Mm, I love pizza," an active listener could say, "I didn't know you liked pizza so much. We'll have to make it again soon." If the child says, "Look what I painted," the active listener might say, "Wow, I see that you mixed the red and yellow paint to make orange." In this last example, the listener makes an objective comment about something observable rather than an assumptive comment about the content of the painting.

4. BE CURIOUS

> *If this (the starry sky) were a sight that could be seen*
> *only once in a century . . . this little headland would be*
> *thronged with spectators. But it can be seen many scores of nights*
> *in any year, and so the lights burned in the cottages*
> *and the inhabitants probably gave not a thought*
> *to the beauty overhead; and because they could see it*
> *almost any night, perhaps they will never see it*

—Rachel Carson, environmentalist and author of *The Sense of Wonder*

Children are great at noticing small details, and encouraging them to look carefully can help tune their attention to moments or objects that get overlooked as they get older. When you're out on a walk or drive, comment on the changing colors of leaves, point out a blimp that moves into your field of vision, or ask your child to point out what he notices.

Anything can be the focus of careful looking. Pay attention to the color of a rising moon, a slow-moving caterpillar, or the smell of low tide. Bring these points of interest to your child's attention and then slow down to marvel at them together. Many adults aren't as good at slowing down as children are, so this last part can be difficult. However, the heightened awareness will help your child develop life skills of wonder and curiosity that are central to forming a creative mind-set.

5. SEE MISTAKES AS GIFTS

I've missed more than nine thousand shots in my career.
I've lost almost three hundred games. Twenty-six times I've been trusted
to take the game-winning shot and missed. I've failed over and over
and over again in my life. And that is why I succeed.

—Michael Jordan, professional American basketball player

Of course none of us wants our children to fail, but small failures can be instructive and helpful as tools for learning about how the world works—and for improving ourselves. Whether we like it or not, life is full of failures: missed birthdays, sunken cakes, dented fenders, skinned knees. These things happen, and there's no way around them. While we can't control the mistakes, we can control the way we think about them. Failure is obviously not our goal, but learning and growth are. So on our way through life's challenges, when we encounter failure in our path toward other opportunities, we can remain open to the lessons it teaches.

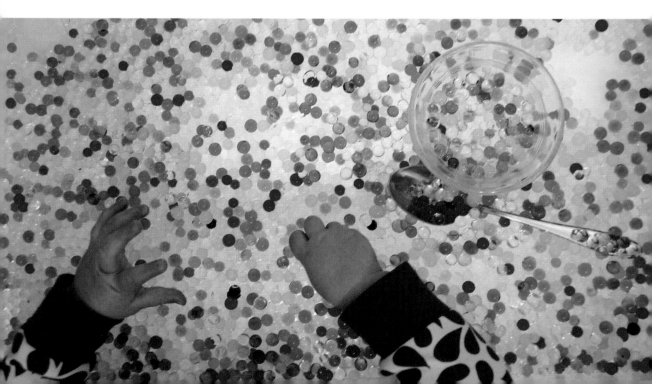

If we encounter mistakes on our route to awesomeness, why not look at them as gifts rather than problems? As gifts, failures become opportunities to learn, improve, describe, and grow, and they can release us from the guilt associated with not being good enough. Carol Dweck, Stanford psychology professor and author of *Mindset: The New Psychology of Success,* says that children are more likely to think creatively if they can view failures as surmountable blunders.

When mistakes happen in my home, we make a point to laugh about them before moving on or trying again. When my kids don't catch a ball, we'll giggle about slippery fingers. When they're unhappy about a drawing, we'll offer a new piece of paper or the opportunity to correct the mistake. If we can help children see these "failures" not as mistakes but as part of a learning process, they'll learn to embrace setbacks as opportunities that can help them improve. They'll also understand that mistakes are part of the natural fabric of life and see them as stepping-stones toward "getting it right."

When a child has trouble getting past the disappointment of a misstep, try asking, "What can we do to change this outcome?" or "How could you do this next time to have it turn out differently?" In a nutshell, look at mistakes as opportunities for growth rather than indications of failure.

6. EMBRACE A GOOD MESS

*Cleaning your house while your kids are still growing up
is like shoveling the walk before it stops snowing.*

—Phyllis Diller, American comedienne

Raise your hand if you enjoy a clean space and feel overwhelmed when messes get out of hand. If your hand is up, you're in good company with a majority of my blog readers. Since I started my blog, I've found that most of the resistance to hands-on projects results from a fear of messes. And glitter. Let it be known that I'm not afraid of either, but I do understand the fear.

It's not the mess per se that many of us dread; it's the time involved in cleaning it up when the fun is done. I have a friend who recently banned paint from her house because her three boys put it everywhere *except* on canvas or paper. I know of another parent who won't allow playdough in her home because her entire house is carpeted. As much as we'd like to encourage our children to help us clean up, their efforts are not always as tenacious as we'd hope.

But if you can, put all this fear aside for a moment and hear the case for making messes. When children are in their creative element, contents will spill from baskets, couches may turn into forts, and baking soda/vinegar concoctions may overflow onto tables. Messes are evidence of creativity, and there's no real way around this. I like to think of messes as battle wounds—they may not be pretty, but they're proof that fun has been had and ideas have been followed. When my kids want to pull out all our sheets to build forts and turn our house into a mini-circus, my first thought is often how I can mitigate the mess. But I try to curb this nay-saying voice with the knowledge that my girls want to test their ideas and follow their curiosity.

You can curb the mess by putting reasonable limits on the experience, but it also helps if you can simply allow messes to happen. The next time your child asks to paint or wants to dump a bag of cotton balls all over the floor, make room for it (and/or take it outside). You may have to contend with a mess, but the creative benefits will surpass this temporary inconvenience. Messes come and go, but creative thinking can be forever.

7. ACCEPT BOREDOM AS A TOOL FOR SELF-DISCOVERY

The cure for boredom is curiosity. There is no cure for curiosity.

—Anonymous

Boredom is awesome. Not boredom itself, but what boredom can lead to. I had the chance to pose a question to Jad Abumrad, founder and cohost of the public radio show *Radiolab,* after a talk he gave on the challenges of telling science-based stories. (If you're not familiar with Jad or his show, suffice to say that he's incredibly creative and that *Radiolab*'s one-of-a-kind quality carries the mark of genius.) When I asked him to reflect on what aspect of his childhood led to his creativity, he paused and said simply, "Boredom."

The first few moments of boredom may feel uncomfortable or seem never-ending, but as we push through them, we have to face our ideas, passions, interests, and curiosity. When you watch a child who's pushed past boredom, you might see her reading, singing to herself, tinkering, inventing stories with small action figures, or building a miniature city. While we do our fair share of facilitated projects, when my daughter tells me she's bored, I simply ask her to find something to do, knowing she'll come up with something amazing that will help her build connections between the disparate thoughts of her growing mind.

Children need unstructured time to follow their curiosity: to imagine, build, experiment, and explore. It can be an hour before dinner or time set aside for imaginative play every weekend—the idea is for a child to face boredom and conquer it. This may seem obvious to some of us, but I think it's worth mentioning because so many of today's children lead highly structured lives, running straight from school to playing sports, practicing an instrument, or taking a dance class.

Look at your schedule and make sure there's time set aside for "nothing." When children have nothing to do, they have to face their own ideas.

8. STEP BACK AND ENJOY THE FLOW

Learning takes place best when young children are engaged and enjoying themselves.
    —Kathy Hirsh-Pasek, *A Mandate for Playful Learning in Preschool*

Have you ever heard the Chinese proverb "Teachers open the door, but you must enter yourself"? Let's open doors and introduce our children to wonderful things, then step back to witness them actively engage in their learning experience with minimal reliance on our instruction or interruption.

Think of a time when you were completely absorbed in an activity (reading, writing, making something, playing a game) to the point that your attention

couldn't be disturbed. Where were you? What were you working on? How much time passed before you came up for air? If you can remember this moment, you'll know what it feels like to be "in the flow."

The acclaimed psychologist Mihaly Csikszentmihalyi introduces us to the concept of flow in his book *Flow: The Psychology of Optimal Experience.* The idea is simply that people are happiest when they're deeply absorbed in an activity. When a child is deeply engaged in a moment, do everything you can to step back and allow the moment to progress without interruption. When my girls are engaged in quiet play, drawing, or writing, I try to refrain from distracting them with conversation or jumping in with questions like, "What are you drawing?" If a child feels like she's constantly under surveillance, she may be less likely to take risks, which diminishes her creativity. If you encourage autonomy, you'll see your child's imagination bloom. So go ahead and pour yourself a cup of tea and enjoy a moment to yourself!

9. SPEND TIME OUTDOORS

> *When I go into the garden with a spade and dig a bed,*
> *I feel such an exhilaration and health that I discover that*
> *I have been defrauding myself all this time in letting others*
> *do for me what I should have done with my own hands.*
>
> —Ralph Waldo Emerson, American essayist and poet

In her book *The Creative Habit,* the dancer and choreographer Twyla Tharp describes Beethoven's creative habit: "Although he was not physically fit, Beethoven would start each day with the same ritual: a morning walk during which he would scribble into a pocket sketchbook the first rough notes of whatever musical idea inevitably entered his head. Having done that, having limbered up his mind and transported himself into his version of a trance zone during the walk, he would return to his room and get to work."

Being active and spending time outdoors are a big part of our lives, but perhaps not the most obvious way to support creative growth. Nature provides children with sensory experiences and heightened personal awareness that they simply can't get indoors. The surprises that nature offers are full of so many opportunities to think in new ways: a deep puddle left behind by the rain, animal tracks that lead beyond the path, the perfect tree for climbing. Moments like these can open a child's mind to make new connections and look at the world with fresh eyes.

10. THINK OF EVERYTHING AS AN EXPERIMENT

Creativity is like scientific research in that it involves doing things that haven't been done before. As such, creative endeavors are essentially experiments, and if they are really unique, you have no idea what will happen.

—Tina Seelig, *inGenius: A Crash Course on Creativity*

The other day, my two-year-old told me she was going to do an experiment. I asked her to tell me what an experiment is, and she said, "An experiment is . . . you know." Hmm. Then she climbed onto my bed, stood tall in the middle of it, allowed her body to topple over, and finally stood up and laughed. Only she knows what was going on in her mind, but the experience made me think of the elements of the scientific method: asking a question (What will happen if I stand in the middle of this bed and allow myself to fall over?); conducting an experiment (toppling over); and analyzing the results (Hey, I landed safely!) to make a discovery.

Anything can be an experiment, and we can listen for questions and observations that can lead to these tests of curiosity. Wondering about the sun could segue into drawing shadows. A fascination with rhyming words might lead to inventing songs. Questions about spiders could lead to hunting for spiderwebs, drawing spiders, and reading books about them. One thing leads to another, and new questions will certainly emerge from each round of discovery and experimentation.

A habit of experimentation is good for many reasons. Experiments teach children that there are multiple ways to approach a problem. When children solve self-designed problems, they learn how to think for themselves. Experiments also remind parents that they are colearners who don't have all the answers. The spirit of experimentation, exploration, and pushing boundaries is at the root of innovative thinking.

While you may still be in the process of setting up your self-serve space or making a list of supplies to stock your shelves, creative thinking doesn't wait for the perfect time or place to begin. Don't allow a lack of "just-right" supplies or confidence as a parent facilitator stop you from jumping into hands-on making and experimentation today. Go ahead and get started, and know that this book will be waiting for you as a ready friend to give you a boost of confidence or nugget of inspiration when you're ready to try something new. If things don't go quite as you expect, that's okay and part of the process—creativity is as much of an experiment for you as it is for your child. And what better way to model a spirit of inquiry and experimentation for your child than by discovering the unknown together.

Experience

DESIGN

There is an energy in the creative process
that belongs in the league of those energies
which can uplift, unify, and harmonize all of us.
This energy, which we call "making," is the
relating of parts to make a new whole.

—Corita Kent and Jan Steward,
Learning by Heart: Teachings to Free the Creative Spirit

I clearly remember the moments when each of my children made their first marks on paper. In each case, I placed the largest sheet of paper I could find on the floor and invited them to draw; one used markers, and the other used crayons. The attention they gave to the invitation was slightly different, but both explored how to hold the drawing tool and experimented with making a variety of strokes. These first marks were a far cry from depicting realism and more about exploring the motion of their arms while making connections between themselves and their surroundings.

The design projects in this first section invite children to use supplies such as markers, crayons, paper, glue, scissors, and paint to create and construct drawings, sketches, and patterns of their own design. These experiences offer children a multitude of gifts that go beyond the simple joy of hands-on making: they can help develop fine motor skills and hand-eye coordination, build creative confidence through the exploration of new materials, and increase visual vocabulary.

OH NO, THAT'S NOT CREATIVE!

Jessica Hoffmann Davis, EdD

What counts as artistic activity? Is it whenever our children lift crayons to paper, voice to song, or body to dance? Or is it when our children do these things in a certain way? Some parents value their children's drawings when they seem to resemble particular objects; they applaud notes that are sung on key and dances that lack mistakes. Others embrace the expressivity of colorful scribbles, the belting out of a song, or the formless leaps of an eager dancer. As a young mother, I fell into the second category, and I was wary of opposing views. I outlawed what I saw as constrictive coloring books and challenged teachers who marked worksheets "wrong" when a child colored outside the lines.

As an art teacher in the 1960s, I applauded three- to five-year-olds who illustrated stories as they told them with quick changes in the direction of live-action finger paint. Their messy images were alive with color and movement, and their crayon drawings were exquisitely expressive. Later in the day, I would shake my head as I watched the older children copy each other's tight and tidy drawings. Sitting in a row, one eight-year-old would make a rainbow, and that rainbow would appear in the drawings of the next child and the next. The flowers were the same—in a row on a green line that counted as grass. Then there would be that house, the all-too-familiar box and triangle.

"What a lack of creativity," I would think. The stereotypical shapes they used—stick figures for people or lines and circles for trees—were for me an anathema to artistic development. I wanted my students to make their own drawings, using colors and figures as vibrant and expressive as they had before they went to school. I wanted them to realize they could create a world with an orange sky and purple ground and that square houses and colored rainbows were not necessary parts of an invented landscape.

While I saw myself at this time as a guardian of creativity, I think now that I was shortsighted in dismissing the personal triumph of careful correctness or the sense of community that is born in a shared vocabulary of visual stereotypes. I believe now that the only time these activities threaten artistic development is when they discourage young artists from carrying on, from going beyond the given they must know to the invented they will create.

The real challenge is keeping alive our children's enjoyment of artistic activity—replete as it may be with emotive images, tidy lines, celebratory copying, original gesture, or conventional steps. Most five-year-olds will agree that they are artists. Most twelve-year-olds will insist that they are not. This is unacceptable.

We cannot predict which of our children will grow up to be professional artists, but we can insist that all of our children gain comfort in employing and reading artistic symbols. We must insist that our schools provide opportunities in the arts throughout our children's education. And it is our responsibility as parents and teachers to celebrate our children's creative exploration, whether the artistic roads they choose are wild with bends or drab with straight and narrow.

CIRCLE GAMES

Circles are all around us, and we're drawn to them as symbols for the sun, faces, people, and other familiar objects. According to Betty Edwards, author of *Drawing on the Right Side of the Brain,* circles are one of the first recognizable shapes that children spontaneously draw. This was true for both of my children, and you may find that it's the same with yours. These five circle exploration games are arranged to follow the progression of a child's developmental readiness, although feel free to try them in any order.

Supplies
- See individual invitations

INVITATIONS

Circle Stickers. Offer your child a sheet of paper and a couple sheets of circle label stickers, which you'll find in the office supply aisle. These easy-to-use stickers make them one of my all-time favorite art materials for children.

Tracing Circles. Offer your child a sheet of paper, a marker or crayon, and a circular object such as a bowl to trace.

Cut a Spiral. Cut a piece of paper into a circle shape that is at least 8" across. Offer your child the paper and some mark-making tools, and invite him to color on the paper. Once the drawing is complete, draw a spiral on the paper, and cut the spiral out. Attach one end to a piece of string and hang the spiral from the ceiling.

Draw on Paper Plates. Offer your child a paper plate and some markers. On your own plate, show him how he can mimic the shape of the circle with various lines, colors, and patterns. Encourage him to invent patterns of his own. You can also try this with round doilies.

Filling Circles. Fill a paper with circles and invite your child to decorate them.

PEEL AND STICK

Children love stickers. We get stickers at the grocery store, office
supply store, and art store. Some of my favorite stickers for children are the
circular and rectangular labels used to price yard-sale items; you can find them in
the office supply aisle. They come in multiple colors and shapes; are easy for small
hands to remove from the paper backing; and are economical for kids who want
to stick, stick, stick. Paper or washi tape can be used with or instead of stickers for
these invitations. Think of sticker art as early-level collage for little hands.

Supplies
- Paper
- Stickers
- Colorful tape

INVITATIONS

Stickers. Sit down with your child, and place a sheet of blank paper and a sheet
of colorful stickers in front of each of you. Start peeling stickers off your sheet and
putting them on your paper—your child will soon catch on and follow your lead.

Older children will enjoy the challenge of inventing patterns with the stickers.

You can encourage this by creating patterns on your own paper. Take this a step further by combining stickers with other items like toothpicks or small scraps of paper.

Tape. Tear or cut small pieces of tape and attach them to the edge of a table, chair, or window ledge. Invite your child to choose the pieces she wants to stick on her paper.

EXPERIMENT

- **Fill a frame.** Draw a frame on a piece of paper and invite your child to fill it with stickers.
- **Invent something.** Offer your child stickers and toothpicks. What will she create?
- **Make a rubbing.** Cover a paper with stickers. Place a blank sheet of paper on top of it and rub along the paper with the edge of a crayon to reveal the pattern of the stickers underneath.
- **Decorate eggs.** Place stickers on eggs before dying them. Remove the stickers after the dye dries.

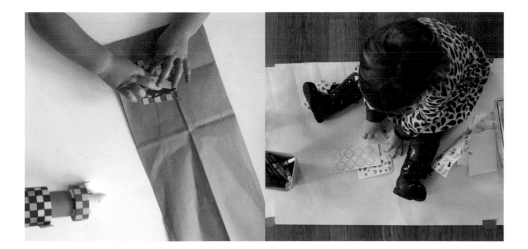

GLUE, GLUE, GLUE

When my girls first handled their own glue bottles, all they wanted to do was squeeze. If this happens to you, be prepared for mountain upon mountain of gooey white glue. Once I explained to them how glue operates as a way to attach things, their love for it grew exponentially.

The oh-so-simple experience of gluing is a winner for developing fine motor skills and also introducing the concept of how one object can connect to another. Children will practice picking up small objects and, as their skills develop, will learn to squeeze the glue on their own. If your child is absolutely new to glue, begin with the first invitation and work forward from there. Food coloring or liquid watercolors can be added to any of these projects for a more painterly effect.

Supplies
- Paper
- White glue
- See individual invitations for specific items

INVITATIONS

Glue Dots. Squeeze small drops of glue all over a sheet of paper for your child. Precut colorful pieces of paper and invite your child to attach the paper to the glue dots. For variety, offer your child a bowl of pom-poms. Gently lift the paper and wave it to demonstrate how the glue helps the papers stick together. (Use your best judgment when introducing young children to small objects.)

Glue Jar and Cotton Swab. Fill a small jar with a few tablespoons of white glue and add a cotton swab to spread the glue on a piece of paper. Offer your child a bowl filled with small pieces of paper or objects such as sequins to attach with the glue.

Glue Bottle. Give your child control of the glue bottle. Invite him to squeeze glue onto the paper and decorate it with small objects such as buttons, beans, seeds, and balls of tissue paper.

FIVE-MINUTE EDIBLE GLUE RECIPE

Supplies
- 1 cup flour
- ⅓ cup sugar
- 1½ cups water
- 1 teaspoon vinegar
- Food coloring or liquid watercolor to color glue (optional)
- Glitter to mix into glue (optional)

Combine the flour and sugar in a saucepan. Add ¾ cup of the water and mix until it forms a thick, smooth paste. Add the rest of the water and mix until smooth. Stir in the vinegar. Cook over medium heat until the paste thickens. Cool. Store in an airtight glass jar or plastic container.

The glue will last up to four days at room temperature or for months if kept in the fridge. When you're ready to use it, scoop some into a bowl and add a brush. If the glue is too thick, thin it with a little water. Note that this glue dries more slowly than store-bought glue.

MARK OUTSIDE THE BOX

My painting does not come from the easel. . . . On the floor I am more at ease.

—Jackson Pollock, American painter

My two-year-old and I visited her dad at work and arrived just before the class he was teaching ended. To keep us occupied, he handed us a few dry erase markers to use on a sliding glass door. My daughter was mesmerized by the opportunity to draw on this huge, seemingly off-limits "canvas." While students who walked by were a little shocked to see a small child drawing all over the university's doors, it was perfectly fine, and she owned this unique experience. Drawing in unusual places helps children see the world as their canvas. Consider new ways to approach mark-making that suit your child's interests. The following invitations will get you started.

Quick tip: Be very clear about which things can be marked up and what's off-limits. You can thank me later.

Supplies
• See individual invitations for specific items

INVITATIONS

Dry Erase Markers on Glass. Find a large window or mirror, offer your child a dry erase marker, and let the fun begin! These markers wipe off easily with a soft cloth. If you have a tricky spot, glass cleaner will take it right off.

Chalk on Fences, Sidewalks, and Trees. Provide your child with chalk and encourage her to draw on sidewalks, trees, and wooden fences. Rain will naturally wash chalk off sidewalks, but trees and fences may need a quick spray with a hose to clean them.

Upside-Down Mark-Making. Tape a large sheet of paper to the underside of a table for an upside-down drawing challenge. Older children may enjoy coupling this experience with learning about how Leonardo da Vinci painted the Sistine Chapel.

Draw Big. Making things big—I mean *really* big—is beyond exciting for many children. Cover a fence or outdoor wall with butcher paper. Or roll a huge expanse of paper out on the driveway or over some grass. We've done this both ways, so why stop with just one approach? Fill a few bowls with washable tempera paint and place some brushes nearby. Invite your child to paint away.

Draw on Cookies. Draw on baked sugar cookies with food-safe pens or paintbrushes dipped in food coloring. You can then eat your creations!

Draw in the Sand. Find sturdy sticks on the beach (or bring your own) and leave scribbles in the wet sand. Watch the waves erase your drawings and start again.

Draw on a Banana. While people have been decorating bananas for years, the artist Phil Hansen popularized this technique by tattooing bananas with pinpricks. To tattoo your own banana, scratch its surface with a toothpick or prick it with a pin. As the exposed skin oxidizes, it will turn brown to reveal your design. If you want to preserve—and eventually eat—your decorated banana, place it in the fridge to keep it fresh.

EXPERIMENT

Talk to your child about how art can happen anywhere and that almost anything can be an art material. To think of other outside-the-box ideas, ask questions like, "What other materials could we draw with? What else can we draw on? Where else could we tape paper for drawing?"

WATERCOLOR EXPLORATION

*Painting is the act of discovery, and you're constantly enlarging
your horizon or finding yourself every time you paint.*

—Romare Beardon, American painter

Watercolors are economical, easy to clean up, and dry quickly, making them one of my favorite art-making supplies. Dry watercolor pans are best for traveling, but when we're at home, I have a supply of liquid watercolor bottles that are in constant use. I prefer liquid watercolors because they're vibrant, they encourage color-mixing tests, and we can use them interchangeably with food coloring in our playdough recipes and liquid experiments. For watercolor paper, I wouldn't recommend anything too expensive, since children are generally more concerned with process than product. However, if you plan to frame the finished work, you could look for a watercolor block at art stores, which will keep water-soaked paper from wrinkling.

Supplies

- Watercolor paper (The more water you add to the paper, the more it will wrinkle. The heavier the paper, the less it will wrinkle.)
- Liquid watercolors or pan of dry watercolors
- Small jars, muffin tin, or ice cube tray to hold the liquid watercolors
- Watercolor brushes (I like round, size 8 brushes, but kid-grade brushes are just fine for beginners.)
- Sturdy jar, half-filled with water
- Rag or paper towels to absorb wet brushes
- Paper tape, optional
- Eyedroppers
- See individual invitations for specific items

INVITATION

Place a piece of paper on the table and the other materials on either side of the paper. My children prefer using their right hands, so I tend to place the paint, brush, water, and rag on the right side of their paper. Tape the paper to the table if it's in danger of moving around. Invite your child to paint.

EXPERIMENTS

Salt and Glue. Salt adds a sparkly, glittery effect when added to watercolor paint. Draw a glue design on a piece of watercolor paper, poster board, or card stock. With an eyedropper, squeeze liquid watercolors on top of the glue design. Sprinkle salt on top. Pour excess salt off when the paint is dry.

Crayon or Oil Pastel Resist. Draw a design on heavy paper with crayons or oil pastels. Encourage your child to press firmly, but not so hard that the crayon or pastel breaks. When the design is complete, paint over it with watercolors. Try this effect with different color combinations. White crayons and pastels appear invisible on white paper but show up clearly with the addition of paint (older children love this trick).

Coffee Filters or Paper Towels. Round, white coffee filters are one of the most economical art materials I know of, and we use them all the time for watercolor mixing experiments and cutting out snowflakes. Place a few coffee filters or paper towels on a large cookie sheet or on a tray with sides. Fill small glass jars or plastic containers with liquid watercolors and eyedroppers. Invite your child to drop the paint onto the filters or towels. As the paper absorbs the color, the results will probably look a lot like tie-dye. Have a long sheet of butcher paper set up to absorb the liquid from the drying filters.

Spray Painting with Watercolors. Fill a spray bottle with water mixed with a few drops of liquid watercolors. Invite your child to spray designs on a large sheet of paper, an old sheet, or a tarp that's been secured to an outdoor wall or fence. You can also place paper flat on the ground, arrange leaves on it, and spray paint over the leaves to create silhouettes.

Blowing Paint. Place a sheet of paper on a tray. Fill small bowls with a 1:1 mixture of water to watercolor paint. Give your child an eyedropper and invite her to draw a little paint into the dropper and drop it on the paper. Blow on the paint through a straw. What happens? To kick this up a notch, offer your child a hair dryer in place of the straw.

WATERCOLOR TECHNIQUES

Tip #1. When you want to clean your brush, place it at the bottom of the water jar and make it "dance" up and down until the brush is clean.

Tip #2. If you have too much water on your brush, dab it on a towel to absorb some of the extra water.

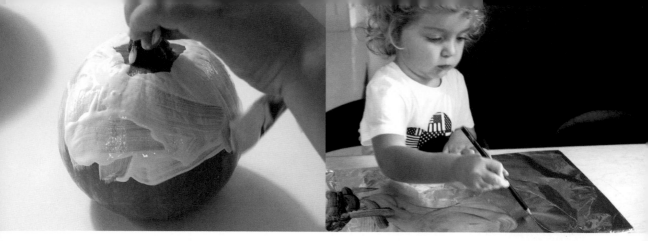

YES, YOU CAN PAINT ON THAT!

At a very early age, I recognized everyday objects as canvases for the imagination. My siblings and I were raised in a creative home where my dad had a knack for designing jewelry and painting rocks (yes, it was the seventies), and my mom's hand-painted Scrabble set came out every few weeks. When children see that anything can be a surface for making art, they're empowered to find the extraordinary in the ordinary.

Supplies
- Objects you don't usually paint on, such as rocks, boxes, leaves, seashells, sticks, dowels, and birdhouses
- Watercolors, tempera paint, or acrylic paints
- Paintbrushes
- Paper-plate palette
- Jar of water
- Rag to wipe brushes and spills

Note: Tempera and watercolors are my favorite childhood paints because they're washable. Acrylic paint is plastic-based (not washable) and a great choice if you're painting on wood, cardboard, foam core, pinecones, and seashells. It sticks to just about everything! Acrylics will stain clothes permanently, so prepare your child and surface appropriately.

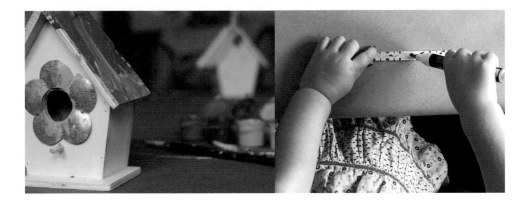

INVITATION

To begin, present your child with a few found or natural objects that he can paint. (Once he has painted his first stick or cereal box, he'll be ready to think about other objects, and then you can seek them out together.) Squeeze small amounts of paint onto a paper plate and invite him to paint the object. You can pour liquid watercolors into small bowls or an ice cube tray.

After you're done painting, you might ask your child one or more of the following questions: What was it like to paint on this object? How was it different from painting on paper? Which do you prefer? What other objects could we paint on?

EXPERIMENT

An alternative that both my kids adore is to "paint" on objects with acrylic paint pens or permanent markers. The mechanics of *drawing* are far different from those of *painting,* and drawing with these tools does wonders for developing fine motor skills and encouraging the process of creating detailed illustrations.

MONOPRINTS

The word *monoprint* comes from the root *mono-,* meaning "one." With this project, children will learn how to make and pull one unique print. There's something magical about the surprise of revealing prints that makes this a favorite from my teaching days. This process can be addicting, and my children often turn into little printing machines when we monoprint.

First, let's talk about printing ink. There are a couple options for this project: tempera paint and water-based printing ink. Tempera paint is nontoxic, washes up easily, and is convenient since there's a good chance you already have it on your supply shelf. Printing ink will not wash out of clothing, and it takes a bit more elbow grease to clean it up, but it's more archival and creates a cleaner, stronger print than tempera paint. If you're printing for pure fun, go ahead and play with tempera paint, but if you're making prints as gifts or something to hang on a wall, water-based printing ink is a good way to go.

Supplies
- Cookie sheet, piece of acrylic, plastic tray, or large ceramic tile
- Brayer, paint roller, or wide paintbrush
- Tempera paint or water-based printing ink
- Printer paper
- Cotton swabs

PREP

This multistep project requires a bit of preparation, but the extra effort is well worth your trouble. Take a few minutes to get everything ready so things will run smoothly.

Printmaking can be messy, so cover your workspace if needed. Clear a large table or area of the floor for drying completed prints.

INVITATION

Place a cookie sheet, brayer, and cotton swabs in front of your child's seat. Place a stack of paper next to the cookie sheet so you can pull multiple prints without too much scrambling around.

Invite your child to squeeze a small amount of tempera paint onto the cookie sheet. Spread the paint around with the brayer. Draw designs in the paint with fingers or a cotton swab.

Place a sheet of paper on the paint. Press the paper down with the palms of your hands. Peel the paper back. Voilà!

Place the print on your cleared space and continue pulling prints until your child runs out of steam.

EXPERIMENT

- Try printing on different types of paper—newspaper, construction paper, origami paper.
- Experiment with different drawing tools such as a paintbrush, pencil, crayon, or feather. How do these marks affect the final print?

BUBBLE PRINTS

If you've ever watched a child try to catch or pop bubbles, play with bubble bath, or blow bubbles in her milk, you've witnessed the fascination that bubbles hold for little people. Now imagine filling a glass with a colorful bubble solution and inviting your child to blow into it. It's a child-ready invitation, right? A few tips on bubble blowing: Since our natural instinct is to inhale when given a straw, practice blowing and exhaling through the straw before you begin. Even with this practice, my three-year-old still sucked up a little paint on her first attempt, which is a great reason to use nontoxic paint. Don't let this deter you—once a child gets the hang of making straw bubbles she may be hooked.

Supplies
- Medium-sized bowl
- Water
- Tempera paint
- Straw
- Dish soap (This works best with standard rather than all-natural dish soaps.)
- Heavy-weight paper

INVITATION

Fill the bowl with a 1:2:2 mixture of water, tempera paint, and dish soap. Mix well. Place a straw in the dish soap mixture and encourage your child to blow until bubbles appear and pop up over the rim of the bowl. Place a sheet of paper on the bubbles to capture a print. Repeat as many times as you like.

EXPERIMENT

- Try blowing softly and then with more force. Which technique creates the best bubbles?
- Set up multiple colors of bubble paint and layer colorful prints on top of each other.
- Make a few batches of bubbles using different kinds of dish soap or bubble bath as the base. Compare the results. Which do you prefer for printing?

DRAWING GAMES

Humans are born playful, and when creative exploration has play at its core, surprising things are bound to happen. These are some of our favorite drawing games. I can't tell you how many times we've begun one of these games only to have it move in a new direction, so consider these ideas as jumping-off points and always be open to inventing your own.

Supplies
- Paper
- Markers or crayons
- See individual invitations and games for specific items

INVITATIONS

Slide Drawing. Roll a long sheet of paper out over a slide and tape it down with masking tape. Offer your child crayons and encourage her to draw as she slides.

Cut Shape Creatures. Cut an abstract, creature-like shape from paper and give it to your child to fill in the details. Provide googly eyes that can be stuck anywhere.

Organic Shape Creatures. Create an organic shape on a piece of paper with a 20" piece of yarn. Trace around the outside of the yarn. Look at the shape for visual clues, then determine how you'll transform it into a creature with eyes, a mouth, horns, arms, wings, and so on.

Connect the Dots. Draw a series of dots that are completely abstract or that represent a two-dimensional symbol (such as a star, a flower, or a house) and invite your child to connect the dots in any way he likes.

Draw Inside of Shapes. Fill a large sheet of paper with a variety of shapes. Pass the paper to your child and invite him to fill in the shapes with lines and/or solid colors.

GAMES

I Draw, You Draw. The first player makes a line, shape, or scribble on a piece of paper and then passes it to the second player to add another mark. The child or adult can go first, and the drawing can be completed in two rounds or go on indefinitely.

Exquisite Corpse for Kids. Fold a paper in half. Draw the bottom (or top) of a creature on one half of the paper, extending the lines a bit into the top section. Fold the paper over so that your child can only see the extended lines and invite him to continue the drawing.

Draw That Word. Draft a list of words that describe things your child likes to draw or that could challenge his drawing skills, then cut the words out. Drop them into a jar. Pull out words at random and take turns illustrating them.

DRAW WHAT YOU SEE

My first observational drawing experience was in an after-school art class when I was about seven years old. We were asked to bring in an object to draw, and I selected a big chocolate bar (probably because it was the only way to get my mom to buy me chocolate). I remember spending three sessions working on that image and believed that my drawing looked exactly like the candy bar. I was so proud.

In hindsight, I see that the experience taught me about persistence and how to render something to scale. Drawing from observation not only makes us better artists, but looking closely at the details of our world helps us notice the nuances of textures, shapes, and forms. Beyond aesthetics, the process of drawing teaches problem solving and visual communication, skills that give us the confidence to articulate our ideas.

Invite your child to draw what he or she sees, which will very likely differ from what you see. The drawing does not have to look like the actual object—what's important is that your child looks closely and pays attention to details. This activity is appropriate for postscribblers, while scribblers will adapt to the challenge in their own way.

Supplies
- Objects to draw such as flowers, candy canes, toys, or a bowl of fruit (see below for object and material combination ideas)
- Drawing tools
- Paper

INVITATION

Choose a simple object, such as a flower or a pumpkin, and place it on a table. Provide your child with paper and drawing tools to capture the spirit of the still life and invite her to draw what she sees.

EXPERIMENTS

Many adults have been conditioned to draw cartoon-like caricatures of objects and people. To break away from this tendency to symbolize objects, one of my favorite challenges is to sit down and draw alongside my children.

To make a fun drawing game of it, begin by drawing the object from memory, and then draw the same object from observation. Compare the two: How are they different? If you were to draw the object again from memory, what might you do differently? Which drawing do you prefer? Which drawing was more difficult for you?

OBJECT AND MATERIAL COMBINATIONS

- Candy canes—black construction paper, silver permanent marker, red and white chalk pastel
- Pumpkins—black construction paper and orange chalk pastels
- Self-portrait—stand-up mirror and drawing or painting materials
- Vase of flowers—white paper and assorted markers or crayons
- Book with pictures that interest your child, such as rockets, airplanes, or farm animals—white paper and assorted markers or crayons

ART DICE

Art dice games are a creative way to jump-start drawing. Every toss of the dice becomes an opportunity to explore and expand art vocabulary, drawing skills, color recognition, and shape identification. If you have any spare blocks lying around, you might want to upcycle them into a set of art dice.

Supplies
- Three wood, paper, or plastic cubes
- Paint, paint pen, or permanent markers
- Paper
- Drawing tools

PREP

Make three dice, one for each of these themes: shapes, lines, and colors. Draw the following on the different faces of the cubes, or use these ideas as a starting point and invent your own:

- Shapes—square, circle, oval, heart, rectangle, and triangle
- Lines—vertical, diagonal, zigzag, wavy, spiral, and dots
- Colors—red, orange, yellow, green, blue, and purple

INVITATION

Set out the dice along with some drawing materials. Invite your child to throw one, two, or all three dice and draw what he sees. If more than one child is participating, players can take turns rolling the dice and drawing symbols on their own pieces of paper.

To set up a collaborative drawing game, players share one piece of paper. Each player rolls the dice and draws his interpretation of the shape/line/color on the paper. He then passes the dice to the next player, who does the same thing. This continues for a set number of turns.

EXPERIMENT

Beyond Art Dice. Instead of art elements, cover your dice with movements, textures, moods, and so on. For example, you could play a texture-finding game by pasting images of things that are bumpy, smooth, spiky, soft, scratchy, and rough. You could explore emotions by making the following faces: surprised, silly, sad, angry, happy, and shy. Or get your bodies moving by covering the sides with these movements: jump, skip, spin, crawl, dance, and roll.

PAINT EXPERIMENTS

Entire books have been written on painting experiments. The things that can be done with paint are endless, and I often look to my children for fresh ideas on how to play with it. Use the experiments described here as starting points; pick the ones that interest you or your child, and see if you can come up with some of your own. It's through the process of play and experimentation that new ideas will emerge and develop.

Supplies
- Paint (Each of these projects starts with tempera paint, but feel free to try other paints too.)
- Paper
- See individual invitations for specific items

INVITATIONS

Tabletop Painting. Cover a table with craft paper or an unfolded brown paper bag. This extra-large canvas offers an appealing alternative to the standard-sized construction or printing paper that most children are accustomed to making marks on. Add to that an offer to use permanent markers and paint, and you have an invitation that will not be scoffed at.

Marble and Egg Rolling. Cut a piece of paper to fit inside the bottom of a disposable casserole tray. Drop a marble or plastic egg in a bit of paint and then place it on the tray. Let it roll it around to create interesting designs.

Make Tracks. Dip the wheels of a small vehicle in a tray of paint, and then pull the vehicle all over a sheet of paper.

Paint with String. Dip string into a plate of paint and then "dance" it on a sheet of paper. Alternatively, fold a piece of paper in half and then open it. Holding one end of a string, dip the other end in the paint. Drop the paint-soaked side on the paper, refold the paper, and pull the string out for a surprising effect.

Paint with Combs. Pull a comb through a tray of paint, then drag it across the paper to create patterns.

Paint with Natural Materials. Dip a stick, pinecone, pine needles, seashells, flowers, and so on in paint or mud and use them like brushes. Ask your child what other natural materials she could paint with.

Paint with Sponges, Straws, and Toilet Paper Rolls. Offer your child a few straws, sponges cut into a variety of shapes, toilet paper rolls, and tempera paint for stamping and dragging paint on paper.

Paint over Stickers. Cover your paper with round or rectangular office stickers. Paint over them with tempera or watercolor paint. Remove the stickers to reveal the blank spaces underneath.

Paint with a Roller. Roll a paint roller or brayer through tempera paint and provide your child with a large piece of paper to roll the paint onto.

Sponge Splat. Take this outdoors! Cut a sponge into four long strips and secure them in the middle with a rubber band or twist tie. Dip all or part of the sponge in a bowl of paint and then toss it onto a large sheet of paper. What happens? Repeat with other colors. Use the same sponge if you don't mind mixing colors together. Otherwise, make a few splat sponges.

PASTE PAPER

The inspiration for this experience came from a workshop I took with the artist and art professor Robert Shreefter when I was in graduate school. Stressed out from our studies, paper writing, and research obligations, my friends and I breathed a sigh of relief as we learned this technique and happily pushed gooey paint all over paper for the better part of a day. At first it felt like an inefficient use of my time, but the process was so much fun and so memorable that I had to reenact it with my kids. At two and four, they both enjoyed a good hour of squishing the paint and pushing paste around to create patterns in the paper with various tools. The goopy paint has a way of warping thinner paper, which makes for a highly successful, process-oriented project, so be sure to use heavy watercolor paper if you'd like to repurpose the finished products as gift cards.

Supplies
- One batch paste (see page 86 for the recipe)
- Large sheets (at least 12" × 18") of heavy-weight paper such as watercolor paper or card stock (You can do this with smaller paper, but larger paper is more manageable and rewarding for this experience.)

- Tools to scratch patterns in the paper, such as scrapers with notches, toy cars, hair combs, or cardboard with notches cut into it
- Paintbrushes
- Spoons to scoop paint
- Roll of butcher paper

PREP

Make up a batch of paste *the day before* you want to work on this project. Consider what colors you'd like for your paste papers. You can make as many colors as you like from one batch—I find that two colors at a time is inspiring without being overwhelming. This is a messy project, so it's a good idea to have a few damp rags ready to wipe the floor or to clean sticky hands. Roll a large piece of butcher paper out on the floor to collect the creations as they're completed. Depending on the weather and thickness of the paste, these will take about a day to dry fully.

INVITATION

Set up small bowls full of colored paste on a covered table. Place a large sheet of paper in front of your child and invite her to scoop a few spoonfuls of paste and drop them on her paper however she likes. Push the paste around with a paintbrush, fingers, or tools.

Once it's dry, you can cut up the paste paper and turn it into cards, book covers, accordion books, and more.

EXPERIMENT

You can have a lot of fun playing around with paste paper. Try overlapping colors for interesting effects and mixing patterns from different tools. Forage through your home for things like forks or combs that will create a variety of new patterns.

PASTE RECIPE

- 8 cups water
- 1 cup cornstarch
- Food coloring, liquid watercolor, or natural dyes (see page 173)

Mix cornstarch and 1 cup of the water in a large mixing bowl until the cornstarch dissolves. Bring the other 7 cups of water to a boil. Remove from the heat; slowly add this to the cornstarch mixture while blending with an electric mixer for about 1 minute. You'll notice that the mixture becomes gelatinous almost instantly. Transfer the paste into a large glass or plastic storage container, cover with a lid or plastic wrap, and store in the fridge or a cool place overnight. The next day the paste will feel a lot like gelatin. Mix it with a spoon until smooth when you're ready to use it. Scoop a few spoonfuls into a smaller bowl and add food coloring, liquid watercolor, or natural food dye. Mix well until you reach the desired shade.

TEN LESSONS THE ARTS TEACH

Elliot Eisner, professor of art education, Stanford University

1. The arts teach children to make good judgments about qualitative relationships. Unlike much of the curriculum in which correct answers and rules prevail, in the arts, it is judgment rather than rules that prevails.
2. The arts teach children that problems can have more than one solution and that questions can have more than one answer.
3. The arts celebrate multiple perspectives. One of their large lessons is that there are many ways to see and interpret the world.
4. The arts teach children that in complex forms of problem solving, purposes are seldom fixed but change with circumstance and opportunity. Learning in the arts requires the ability and willingness to surrender to the unanticipated possibilities of the work as it unfolds.
5. The arts vividly illustrate the fact that neither words in their literal form nor numbers exhaust what we can know. The limits of our language do not define the limits of our cognition.
6. The arts teach students that small differences can have large effects. The arts traffic in subtleties.
7. The arts teach students to think through and within a material. All art forms employ some means through which images become real.
8. The arts help children learn to say what cannot be said. When children are invited to disclose what a work of art helps them feel, they must reach into their poetic capacities to find the words that will do the job.
9. The arts enable us to have experiences we can have from no other source and through such experiences to discover the range and variety of what we are capable of feeling.
10. The arts' position in the school curriculum symbolizes to the young what adults believe is important.

MARBLEIZED PAPER
WITH PAINT AND OIL

The process of making marbleized paper can be enchanting. Each piece is different from the last—and oohs and aahs will fill the air as you pull little pieces of colorful paper away from the paint. The magic of this experience lies in the relationship between oil and water: their inability to mix permanently enables watercolor-and-oil shapes to appear on the surface of the water. The paper captures these shapes before they eventually separate. This classic art-meets-alchemy project is fun for budding artists and minichemists alike.

Supplies
- Vegetable oil
- Liquid watercolors or food coloring
- Small bowls and spoons for mixing paint
- Shallow pan, such as a cake pan or casserole dish (The size and shape doesn't matter.)
- Eyedroppers
- Forks or popsicle sticks to stir the paint
- Watercolor paper, cut to fit inside the pan
- Butcher paper to dry completed work

PREP

For each color, mix ½ teaspoon of oil with 1 teaspoon of liquid watercolor in a small bowl. Stir well until the paint and oil appear to be one. Despite your best mixing, the oil and watercolor will continue to separate, so be sure to continue mixing throughout this project.

INVITATION

Fill the bottom of the pan with about half an inch of water—not too much or your paper will sink to the bottom. Invite your child to squeeze some of the oil-paint mixture into an eyedropper and drop dots of paint into the pan. Swirl the color around with a fork or popsicle stick.

Place a piece of paper on top of the water; fish it out once it's picked up some color. Have a paper-covered table ready to absorb these oily, marbleized creations.

EXPERIMENT

- Try using different kinds of paper. Which works best?
- Test different color combinations.
- Draw on the paper with a permanent marker before dipping it in the paint.

PLEXIGLAS PAINTING

Everything that is new or uncommon raises a pleasure in the imagination, because it fills the soul with an agreeable surprise, gratifies its curiosity, and gives it an idea of which it was not before possessed.

—Joseph Addison, eighteenth-century British playwright

Back when I worked at the San Jose Museum of Art, one of our curators installed an amazing video of Picasso painting on glass. The piece of glass was held upright like an easel, and the filmmaker set up his camera on the opposite side so that he could capture Picasso painting in reverse. The magic of this technique stuck with me, and I wanted to re-create it with my kids.

Not just any paint will stick to Plexiglas. While watercolor paint floats off and tempera flakes, acrylic paint is perfect for the job. When I ran my kids through this experiment, they both enjoyed painting on the acrylic sheets. Once the painting was dry, my four-year-old wasn't interested in adding permanent marker to her creation—I think it was simply "done" in her eyes. My two-year-old, on the other hand, seemed to enjoy the drawing part of this experience more than the painting. Remember that all children gravitate to different things, and it helps if you can go with the flow.

Supplies

- Plexiglas, policarbonate, polystyrene, or acrylic sheets, any size (We've used 8" × 10" and 11" × 14", which can be found at hardware stores.)
- Acrylic paint
- Paper-plate palette
- Synthetic paintbrushes for acrylic paint
- Butcher paper to cover your table (Acrylic paint can be hard to remove.)
- Smock or apron
- Jar of water to clean brushes
- Rag or baby wipes for cleanup
- Rag to absorb brush water
- Permanent markers
- Swiss Clips or Gallery Clips

PREP

Acrylic paint won't come out of clothes and can be hard to wash off furniture, so, first things first: Cover your table, put on smocks, and have water and rags handy for easy cleanup. Remove the protective covering from one side of the Plexiglas. Keep the other side covered, if possible, to keep it clean while painting. It's no big deal if paint does get on the other side as it will peel off pretty easily. Squeeze small dollops of paint onto the paper plate.

INVITATION

Invite your child to paint on the Plexiglas surface. Just as the filmmaker captured Picasso's work-in-progress, offer to lift up the Plexiglas and spin it around so your child can see her work in progress. Let her keep painting until she's done.

Acrylic paint dries quickly. If the paint is thin, it can be dry in half an hour. Thicker paint will take longer. When the painting is dry, turn the Plexiglas over and add details to the other side with permanent markers. Attach Swiss Clips or Gallery Clips to the edges of the Plexiglas picture so it can be displayed.

FOAM PLATE RELIEF PRINTS

When an image is carved into the surface of a styrofoam plate, it creates a depression called a relief. If ink is then rolled over the plate's surface, it will cover everything but the relief. When an artist places paper on top of the inked plate, ink from the surface adheres to the paper, leaving the depressed spaces blank. These *relief prints* are different from monoprints (see page 68), because here you can make multiple prints from one plate. Whenever we make foam plate prints, I precut card stock or fancy paper because the finished results often make nice birthday or holiday cards.

Supplies
- Dull pencil
- Brayer
- Squishy, flat foam plate (You can find these in art stores or cut the flat part out of clean, thin foam meat/vegetable trays or foam dinner plates.)
- Tempera paint or printing ink
- Printer paper, card stock, or decorative paper
- Cookie sheet, piece of acrylic, or flat plastic tray

Note: Use water-based printer's ink instead of tempera for an image that will last indefinitely. Printer's ink is not washable, so cover your space and child accordingly. Because of its nonwashable quality, it is also a great medium to use if you'd like to print images on clothing.

INVITATION

This project has many steps, and it helps to break them down for young children. First, offer your child a plate and a dull pencil and invite him to draw on the foam. Press hard enough that it makes a groove but not so hard that the pencil pokes holes through the plate. Once the design is done, squeeze a few dollops of ink or paint onto your cookie sheet and roll it out evenly with a brayer. Err on the side of less ink—you can always add more if you don't have enough. Roll the inked brayer over the plate and set the brayer aside. Gently place a piece of paper over the inked plate and press it down lightly with the palm of your hand. Gently peel the paper back to reveal your print.

EXPERIMENT

- Make relief designs with different tools such as skewers, forks, or straws.
- Make prints with more than one color of paint or ink.
- Print on different types of paper such as newspaper, wrapping paper, or envelopes.
- Make a playdough relief print by flattening a piece of playdough and carving into it with a pencil.

COLLAGE PAINTING

This project is a culmination of everything that's come before. There's some gluing, painting, sprinkling, and cutting. Everything will go on top of a piece of wood or prestretched canvas, making this an archival artwork that can be a treasured for many years to come. The binding ingredient here is the acrylic medium, which comes in either a glossy or a matte finish. Acrylic medium dries clear and will have a nice, smooth feel when dry. I love this stuff and have found that kids do too. Even though it's clear, keep in mind that it won't wash out of clothes or off some table surfaces, so cover up everything you don't want take a chance with.

Supplies
- Acrylic gloss or matte medium (I prefer the glossy finish, but both are fantastic and can be found online or in any art store.)
- Base, such as wood, linoleum, cardboard, or some other sturdy surface to hold the collage
- Paintbrush
- Collage paper, such as aluminum foil, tissue paper, or wrapping paper scraps cut into small pieces
- Scissors

- Bowl for the medium
- Glitter (optional)

INVITATION

If your child likes to cut paper, invite her to chop the paper into bits and pieces. Meanwhile, place the base and the bowl full of acrylic medium in front of her. Invite her to dip the brush into the acrylic, paint it on the base, and attach the collage paper however she likes: flat, crumpled, layered, or rolled up in balls. If you want to put pieces on top of each other, simply paint on more acrylic.

EXPERIMENT

- Sprinkle glitter into the acrylic medium and paint it on the base.
- Add a few drops of liquid watercolor to the acrylic medium to tint it. Work with more than one bowl of tinted medium, and layer the colors.
- Layer pieces of tissue paper on top of each other or other papers and watch as the transparency blends with what lies beneath it.

FINDING YOUR FIVE-YEAR-OLD SELF IN THE ART MUSEUM

Margie Maynard

It is pure joy I am giving you. Look until you see.
—Constantin Brancusi, twentieth-century Romanian sculptor

Throughout my long career as an art museum educator, seeing young children engage with art in the galleries is an experience that never gets old. As adults, we sometimes have trouble seeing the forest for the trees when we visit art museums, but quite often, it is the child in the group who seems to enjoy the most immediate and strongest connections with works of art. On many occasions, I have witnessed children making a bee-line to a painting or sculpture to hug it (or try to), kiss it, talk to it, and then talk to themselves about it. In short, children in museums are very much themselves—exploring and discovering very actively with their eyes and bodies and using their imaginations and emerging independence to make sense of their experiences. When we consider that these are all the same activities that go into *making* art, children's responses are not only age appropriate, but something for adults to envy and learn from.

Unfortunately, running and touching are still officially off-limits in the museum, but nevertheless, the excitement and joy that young children naturally express can teach us all some important tips about looking at art as a family.

Find inspiration, not intimidation. The rarified environment of a museum setting often suggests to parents that everyone must be on their best behavior, but that doesn't mean you can't relax and genuinely enjoy it with your child. Get down on the floor with her, or lift him up to your eye level so you can experience not only the art but the architecture, the light, and the big open spaces together. Speaking of big open spaces . . . Just like a large green lawn beckons your child to run and play, an expansive museum lobby or gallery will inspire her to do the same. That's right, the museum itself is responsible for your child's "bad" behavior. So keep that in mind and take her hand when you enter.

Be yourself at the art museum. Take a cue from your child and give yourself permission to lose your inhibitions and react honestly and openly to the art you encounter. When was the last time you shed a tear or laughed out loud in a museum?

Don't overanalyze the art. Actually, don't analyze it at all. Again, follow your child's lead, invite a spontaneous response, and enjoy the intangibles. Deriving meaning is not always essential to enjoying or reaping something beneficial from art.

Use your entire body when "looking" at art. It's not uncommon to see parents point out something they see to their kids, but why not have a little more fun, raise both arms over your head, and use your whole body to follow the movement of some long, curvy lines in a painting or try mimicking a pose you observe in an artwork. This is where losing your inhibitions (mentioned previously) really comes in handy.

Tell a story. At the risk of contradicting my earlier advice, storytelling is, after all, a principal motivation for art making—not to mention it's something children love. Take turns telling lines of a narrative for a work of art based on what you see. Depending on the age and size of your group and the amount of detail in the artwork, it could result in a long, rambling tale or a short one-liner that captures everything. This usually works best with artworks depicting people or animals, but with older children, you can also have fun with abstraction. One of my all-time favorite responses to a minimalist painting of short, dark, diagonal lines arranged on a white ground came from a seven-year-old visitor who wrote on a comment card, "It looks like my sister is driving me crazy."

I want parents to keep this in mind: our children may be artists' only audience to possess the capacity to approach art with a completely open and discovery-driven consciousness. Once we get all "schooled up," most of us lose the ability to just let light in through our eyes and wonder about what's in front of us. Meanwhile, artists strive their entire careers to tap into their five-year-old minds, when everything was new and anything was possible. It's where the joy of creativity lives.

BUILD

The three-dimensional projects in this section go hand in hand with the two-dimensional design projects we just explored. Some children may gravitate more toward mark-making and others toward building—or they may alternate between phases of interest in both.

A good place for a young builder to begin is with a simple set of building blocks, along with a few vehicles and action figures. Loose parts like these are rich with possibilities and toys you'll most likely never regret owning. The simplicity of blocks means they're sometimes overlooked in a sea of high-tech, blinking, and moving toys. However, their open-ended nature makes them a cornerstone of building, inventing, and small-world play.

As we'll learn in Susan MacKay's essay, loose parts are central to imaginative play and construction. Before you go out and invest in a lot of expensive toys, take stock of the small objects you already have and consider how they might become part of your building kit.

Building requires the skills of an engineer, a contractor, and an artist. When children have the opportunity to construct with tape and tubes or build towers out of straws, they learn how to pose and solve problems, make judgments about the construction process, and come to an understanding of how to put things together. Through this process, they will envision, plan, and invent three-dimensional sculptures and decide how to build strong structures.

THE VALUE OF LOOSE PARTS

An Interview with Susan Harris MacKay,
director of education and the Center for Children's
Learning, Portland Children's Museum and Opal School

Q: If you had to identify one element in your space or curriculum that supports creative thinking, what would it be?

A: A critical component of an environment that supports creativity is an availability of "loose parts." At Opal School, these include all shapes and sizes of blocks; collections of found objects; natural materials such as rocks, sticks, pinecones, and shells; and small treasures of every imaginable kind. These are sorted and presented on open shelving in an aesthetically pleasing way and within the children's reach. Collections of found objects might be sorted by color in small, clear containers and then set in a window to catch the sun. Natural materials might be sorted into baskets. Organizing the materials in beautiful ways inspires and invites the imagination to play.

As the children learn to value these materials and to keep them organized, the teachers begin to offer more and more. There is a correlation between the variety and quantity of materials and the opportunity to stretch creative capacities. Material-rich environments provide endless, sensory-rich possibilities for invention, imagination, recombination, revision, collaboration, and language development.

The one material you likely won't find in and among our loose parts collection is glue. We want the children to learn to be comfortable with the impermanence of their ideas. When they create a structure, collage, or image that holds a story they are particularly fond of, they'll ask us to take a picture. The picture will hold their image, and they are happy to return to their fluid flow of ideas, often in collaboration with peers.

Q: How do you encourage the children to play with loose parts?

A: We invite children to engage with these materials in a variety of ways. We often ask questions that stoke the imagination, such as, "What can these materials do?" or "What do you notice about these materials?" Often, these questions are the most powerful because they let the children know our expectation that they are capable meaning-makers and inventors of ideas. These questions also communicate that we are interested and listening.

In the exhibits of the Children's Museum, we have learned to offer a mix of loose parts—some that are more familiar, like blocks, with some that are less familiar, like film canisters. We intentionally create a scarcity of the familiar, which tends to force our visitors to see possibilities in the unfamiliar. It is amazing to see the surprise and delight in the new ideas that are born from unexpected combinations.

Q: What ideas do you have for parents who want to bring the richness of a loose parts experience into their home?

A: Make a collection. You could take a walk with your child and collect natural materials you find, or you could surprise your child by sorting and organizing some interesting things from junk drawers or the craft store. Goodwill or other thrift stores become fun places for treasure hunting when you are looking for materials and containers to organize them.

Start small. A small number of items in the collection will help your child to value the organization of the collection as much as you do. As you learn together about the possibilities, you can add more.

Organize the collection in a way that pleases you. Is there an open shelf or a windowsill it can inhabit? A mirror underneath or behind adds additional magic and interest.

Offer this collection alongside other toys your child already loves to play with. Connecting the new with the known helps build robust brain wiring that your child can count on for a lifetime. You can design an interesting provocation by creating something that you know will inspire or intrigue your child. Model possibilities for him. Create a surprise and then enjoy some playtime together.

Observe your child at play. Photograph her creations. Write down the stories she tells when she plays. All of these adult behaviors communicate to children that their ideas matter to us.

Have fun exploring playfully together!

GUMDROP STRUCTURES

This was the most popular activity ever when my kids were two to three years old. And once we got past the eating-a-gumdrop business, the project mesmerized them and taught them how to build strong structures. It can also be fun to compare the different structural qualities of gumdrops, marshmallows, playdough, and grapes. After doing this project with just gumdrops, add another material and see where your child's vote goes. (In case you're wondering, my kids still vote for gumdrops.)

Fashioning structures out of gumdrops and toothpicks helps children understand one of the basics of structural engineering: triangles are structurally stronger than squares, because triangles will always keep their shape, while squares can be bent into parallelograms. Bonus: the process of attaching toothpicks to the "sticking" materials builds fine motor skills and hand-eye coordination. Use caution if you're using gumdrops, grapes, or marshmallows. When eaten by small children, these objects can pose a serious choking hazard.

Supplies
- Gumdrops, grapes, marshmallows, or playdough
- Toothpicks

INVITATION

Place the gumdrops and toothpicks in two separate bowls and invite your child to build things. After the child has played with the materials for a little while, you can demonstrate how to form them into square and triangular shapes. Play with the two shapes and ask the child to share which one he thinks is stronger.

EXPERIMENT

- Challenge your child to use these materials in new ways and ask questions that encourage further exploration. For example: Can you use all the gumdrops to build one structure? How tall can you build a structure before it topples over? What will make this structure stronger without wobbling? What other materials could we use to build a toothpick structure?
- Replace the toothpicks with dry spaghetti. Spaghetti isn't as strong, but because it can be broken, your child will have the opportunity to make structures of different sizes and add a diagonal "crossbeam" into squares that will turn them into two triangles and make them stronger.

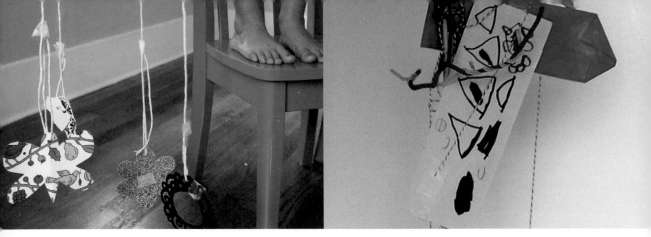

HANGING STRUCTURES

When my oldest daughter was three, she was obsessed with
hanging things from our ceiling. Nothing was off-limits in her mobile-making
madness. Part of the fun in this project lies in selecting the right spot for your
structure: a tree branch in view of the kitchen window, an S-hook in the middle
of a playroom, or a crossbeam on a bunk bed. Show your child how to make a
simple mobile, then step back to see what she does with this new information.

Supplies
- Straws, chopsticks, wooden skewers, or twigs
- Tape, clear or paper
- String
- Heavy-weight paper, such as card stock
- Markers
- Scissors
- Hole punch

PREP

Prepare the hanging structure for a simple mobile by crossing two sticks in the
middle to make an X. Secure the sticks in the middle with tape, string, or a pipe

cleaner. Tie another piece of string at the cross section to be attached to a ceiling hook or taped to the ceiling.

INVITATION

Show your child how she can attach paper shapes to the mobile by dangling them from a string. Work together to make and decorate shapes, punch holes, and hang the shapes from the mobile.

EXPERIMENT

- Once your child has begun making mobiles, free access to a basket of clean recyclables could provide her with a wealth of inspiration. Punching holes in yogurt containers, taping a series of cardboard rolls together, and cutting shapes from cereal boxes are just some of the options. What other materials could you build a mobile from? Paper cups? Egg cartons? Pipe cleaners?
- If you're willing to temporarily part with some of your flatware (or better yet, find some cast-off silverware at a second-hand shop), you can string up a wind chime from forks and spoons.
- Create a mobile that spans from the ceiling to the floor.
- Make a mobile out of natural materials. Hang it in a public place and check it frequently for interaction.

STRAW ROCKETS

I am an adviser for the kids' creative activity kit company Kiwi Crate, and during one of their in-house kid-testing sessions, my three-year-old learned how to make a simple straw rocket. In the process of making a fleet of rockets, she tested a variety of straws and paper. The results: She learned that the shape and size of her rocket affected its speed and the distance the rocket traveled related to the direction and force of the launch. There are plenty of store-bought rockets and launchers out there, but children take a lot of pride in making their own collection of rockets, and these are beyond simple to make.

Supplies
- Straws (We prefer the wide variety that are made for milkshakes and bubble tea.)
- Copy paper, half the length of the straw and about 3" wide
- Circular piece of paper that will become the nose of the rocket, approximately 1.5" in diameter, with a slit cut along the radius of the circle
- Clear tape
- Scissors

INVITATION

Set out the materials and show your child how to make a straw rocket. First, roll the paper loosely around the straw and tape the paper at the top and bottom so it stays together.

Curl the base of the circle around from the slit until it becomes a cone shape. Secure it with tape and attach it to the end of the paper cylinder with two more pieces of tape. This is the nose of the rocket. It may not look elegant, but it should do the trick.

Place the paper rocket on one end of the straw and blow through the other end. Repeat. Laugh. Repeat. Make more rockets, because everyone will need (and want) one.

EXPERIMENTS

- Collect a variety of straws and tubes that can act as rocket launchers, and build rockets to fit them. Take them outside to test and compare the results. Which are easiest to use? Which launch the farthest? Why?
- Build rockets with different weights of paper. Which work better? Why?
- Tape triangle-shaped rocket fins to the tail of your rocket. How does this help or hinder the distance your rocket can travel?

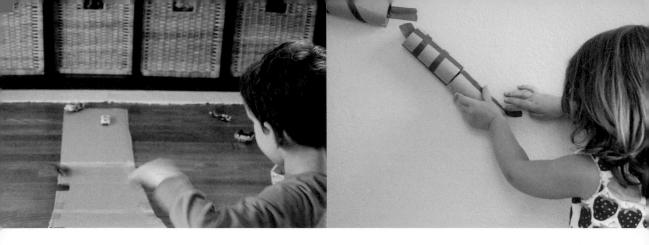

MARBLE RUNS:
RAMPS AND GRAVITY

My children are obsessed with marble runs. We own a store-bought marble run kit that comes out at least once a week, but we've also had enormous fun with our own DIY versions made from cardboard tubes, pool noodles, and storm gutters. Children learn through play, and what better way to learn about gravity, building, and momentum than by engineering wild structures while rolling balls down ramps?

There are many ways to explore ramps and gravity, so I'll share four tried-and-true activities that work for us. Try one or all of them!

Supplies
- See individual invitations for specific items

INVITATIONS

Ramps and Cars. Cut heavy cardboard into long strips, or gather some wood planks. Invite your child to use these materials to build ramps off of a coffee table, shelves, chairs, or other objects, and drive small toy vehicles down them.

Marble Run. Gather some toilet paper and paper towel rolls, painter's tape, and marbles. Find an open wall or sliding glass door that you can tape the cardboard tubes to. Tilt the tubes to create ramps and tape them in place. Test and experiment until you find a flow that works. Challenge your child to create a supertall marble run or a long and winding marble run. This experience is filled with lots of trial and error that teaches iteration and how to overcome small failures.

Gutters and Balls. This is a fun one to take outdoors. Offer your child a few pool noodles that have been sliced in half (vertically) or storm gutters and invite him to prop them against various surfaces to create ramps. How fast can he get marbles, balls, or toy cars to scoot down the ramps?

Water Wall. Build a water wall (see photo on page 42) from recycled plastic bottles, funnels, plastic tubes, and sand pails. Screw the objects directly to a fence or attach them to a pegboard with zip ties or wire. To allow water to run through the water maze, cut holes in the bottles and pails with a utility knife.

EXPERIMENT

- What other objects could you use to build ramps and explore gravity? A pile of books? Three-ring binders?
- Set up two side-by-side gutters and predict which balls will race down the fastest.
- Make predictions about how far or fast a ball will roll.
- Research the American sculptor and inventor Rube Goldberg, and look for video examples of his chain-reaction machines. These are fascinating! Brainstorm how you could make a similar machine at home.

PAPER HOUSES

These houses are made from three simple sheets of paper, and the fun comes with decorating and designing the look of the houses. I've seen both preschoolers and adults meet these parameters with enthusiasm, creating homes that speak volumes about their worldviews or life experiences. This project will encourage children to tap into their imaginations. While considering what they want to place in the room, they'll problem solve, test out ideas, construct, and reflect on the process of making something three-dimensional.

Supplies
- Three sheets of construction paper or card stock
- Recyclables or household goods that can be repurposed for décor and furniture, such as toilet paper rolls, cereal boxes, foam, cotton balls, wallpaper scraps, wrapping paper, fabric pieces, buttons, sequins, and aluminum foil
- Low-heat glue gun with glue sticks
- Scissors
- Permanent nontoxic markers
- Tape
- White glue

PREP

To begin, you will need three pieces of paper, scissors, and a glue bottle or tape. Fold one piece of paper in half and then in half again. Firmly press the creases down. Unfold the paper and place it in front of you horizontally; you should see four quadrants. Cut along the bottom vertical crease, stopping at the middle point. Hold the bottom right quadrant with your right hand and the bottom left quadrant with your left hand. Bring your hands together and overlap the bottom left quadrant over the bottom right quadrant. Secure the two quadrants with glue or tape. Your paper should now look like an open room with two walls.

Repeat these steps for the other two pieces of paper. Two folded pieces of paper will be the room, and the third piece will be the roof. Fit two of the glued pieces together to form a room with three walls and glue them in place. Rest the third folded paper on top of the room and either glue it in place or secure it with tape. If you'd like, you can trim the walls to make the lengths match the size of the roof.

INVITATION

Once the base is built, ask your child to think about what she wants to include in her home. You might want to look around the room you're in or take a tour of your home to gather some ideas. Will there be a window, bed, carpet, or art for the walls? Work together to troubleshoot, build miniature furniture, and turn the blank canvas into a space station, bedroom, or school.

EXPERIMENT

- Build a small town with multiple buildings.
- Make a house from a cereal box. Paint it with acrylic paint and draw designs on it with a permanent marker.
- Cut into the walls to make windows.
- Glue your house to a large piece of cardboard and create an outdoor environment.

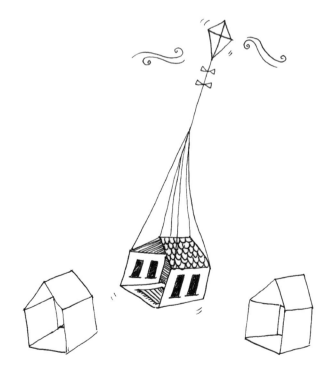

SCRAP BUILDING

Constructing with found materials encourages resourcefulness and helps children recognize everyday objects as sources of building inspiration. It's easy to find and collect objects for scrap building: egg cartons, coffee filters, string, wood scraps, and fabric remnants accumulate in most homes and can be stored in a basket or box in your studio area. Natural materials such as pinecones and twigs are also fun to collect for projects like this and don't cost a thing. If and when structures get really big, children will be challenged to find ways to make their structure stable. This can also be a great opportunity for teamwork and coordination skills.

Supplies
- Recyclables that can be stacked and combined, including boxes, egg cartons, oatmeal containers, milk cartons, plastic bottles, yarn, and wrapping paper
- Low-heat glue gun
- Treasures, such as feathers, sequins, and googly eyes
- Tape dispenser
- Glue
- Stapler

- Markers
- Scissors
- Wastebasket, positioned nearby

PREP

Collect a variety of materials. You may want to work with your child to start an inventor's kit (a treasure box full of recyclables), and add to it over the course of a week or two. When you have a nice selection, you're ready to begin.

INVITATION

Sit down with your child and show him how to put materials together with the glue gun, tape, stapler, and glue. White glue, tape, and staples are best for attaching pieces of paper and lightweight objects, and the glue gun is handy for joining heavier objects. Play around with your own set of materials as a way to inspire your child and prime him with some ideas for exploration. Once he gets the hang of gluing or taping found images and treasures, you may want to bring in more materials.

EXPERIMENTS

- Challenge your child to create a tall structure that won't topple over.
- Make a sculpture that represents something such as a robot, person, car, or machine.
- Offer your child a basket full of materials and challenge him to use everything in the basket.
- Make something from multiples of one object. Place a hundred popsicles sticks or a few dozen paper cups, along with a roll of tape, on a table and have yourself a building party. This project relies on just two things: the object and connectors; for example, wood scraps and white glue, drinking straws and tape, or corks and a low-heat glue gun. Ask your child, "What

could you create with these items?" Be open to any interpretation of the materials. He may use just a few pieces, he may use them all, or he may want to incorporate other materials into his invention.

LOW-HEAT GLUE GUN TIPS

With clear instructions and supervision, children are capable of using a glue gun. I prefer the "low-heat" style—they work well, and I retain my peace of mind just in case my kids touch that taboo metal tip. Point out to your children how hot the tip is and explain that it's not for touching. The idea isn't to fill them with fear but to give them a healthy respect for this useful tool. A demonstration of how it can secure one object to another will help them see its potential.

ROPES AND PULLEYS

Pulleys are simple machines made of a rope that slides around a grooved wheel. They help people raise or lower objects easily by reducing the amount of effort needed to lift them. Pulleys are sometimes referred to as a block (the wheel) and tackle (the rope). Show your child how to use a pulley with some of the tips here, then keep a basket of pulleys and ropes in your garden or construction area as a way to encourage your child to invent ways to use this simple machine.

Supplies
- Pulley
- Utility hook or screws to secure the pulley to a base object
- Spring link (carabiner)
- Rope (a width that fits the pulley)
- Object to hoist, such as a handled basket or bucket

PREP

Talk with your child about pulleys and how people use them to make the work of lifting objects easier. Pulleys can be found on elevators, flagpoles, sailboats, cranes at construction sites and in shipping yards, clotheslines, tow trucks, weight machines, and water wells. There are three basic parts to remember: there's the pulley itself, the force (effort exerted when pulling the rope), and the load (the object being lifted).

Install a utility hook in a wall, fence, or tree. Attach the pulley's hook to the utility hook, run rope through the pulley, and attach one end of the rope to a load such as a bucket of dirt. Demonstrate for your child how the pulley can help hoist objects up and down.

INVITATIONS

Lift a Heavy Object. Offer your child some objects to play with in conjunction with the pulley. For example, ask her if she can lift a bucket of heavy apples without the assistance of the pulley. After struggling with it, attach the bucket of apples to the rope that's connected with the pulley and then invite her to lift the bucket again. Does this make the lifting easier?

Move an Object. This one is like a zip line. Run a piece of rope across a room or garden by connecting it to two hooks. Remove one end of the rope and run a pulley through it upside down, with the hook facing the ground. Reattach the end of the rope to the hook. You should now have a pulley that runs freely along the rope line. Attach a second piece of rope or carabiner to the hook that extends from the pulley. You can now connect a bucket or basket to the rope and run the object along the rope line.

CD SPINNER

This homemade top is one of the best-loved toys in our collection, and from a teaching standpoint I love it for teaching my children about centripetal force and the magic of mixing lines and colors through motion. When the top spins, it's pulled inward by centripetal force, which keeps it upright until friction makes it topple over. While you don't have to get technical about the force behind the spinning, it's interesting to note that *centripetal* means "center seeking," indicating that the force is directed to the middle of the circle.

Supplies
- CD
- Card stock or other heavyweight paper
- Markers
- Glass marble
- Low-heat glue gun
- Plastic water bottle cap
- Scissors
- White glue

INVITATION

Place the CD on top of the card stock and trace around it. Remove the CD and invite your child to color inside the drawn circle. Cut the drawing from the paper and glue it with white glue to one side of the CD. Glue the plastic water bottle cap to the center of the card stock with hot glue. Glue the marble to the other side with hot glue. Spin it!

EXPERIMENT

Not only are these tops fun to make and spin, but they also give children an opportunity to learn about the visual phenomena of optical illusions. An optical illusion is a trick of the eye, where the mind sees an image that differs from what's actually there. You can create different optical illusions by making a variety of patterns on multiple tops. For example, draw a black spiral from the center to the outside of the circle, or cover the paper with yellow and red lines or dots. How do the visuals of these patterns change once the top is in motion?

DOES IT FLOAT?

To solve a problem is to create new problems, new knowledge immediately reveals new areas of ignorance and the need for new experiments.

—Sir George Porter, British chemist

My sister is a professional sailor and has worked on container ships that transport cars, equipment, and train containers between California and various ports throughout Asia. If you haven't seen a container ship up close, they are gigantic! When I see these boats sail under the Golden Gate Bridge, I'm always struck by the quantity of containers that can be stacked on the ship's deck without sinking the boat.

The science behind this bit of wonderment is a principle of flotation, which states that a boat will float when it displaces water equal to its own weight. In essence, the heavier a ship is, the wider it must be in order to stay afloat. A narrow hull and a heavy load are a sure way to quickly sink a ship to the ocean floor. Even though I know that little fact, it's still hard to fathom. But what better way to explore this principle with kids than through a series of tests that explore how boats float and what could make them sink?

Supplies

- Large plastic storage box, kiddie pool, water table, or bathtub filled with water
- Objects to build a boat (This list is just for inspiration—use what you have.)

  - Plastic bowl
  - Plastic food storage container
  - Plastic cup
  - Skewers
  - Duct tape
  - Rubber bands
  - Scissors

  - Exacto knife (for adults only!)
  - Aluminum foil
  - Styrofoam
  - Wood scraps
  - Hot glue gun
  - Straws
  - Styrofoam egg carton

INVITATION

Begin with a discussion about boats. Ask your child some guiding questions, such as, "What do you know about boats? Where have you seen boats? What do you remember about them?" Offer your child the challenge of building a boat that floats, using any of the materials in the preceding list.

Once you've made some boats, discuss which ones work best. Ask your child, "What helps the boats float? What makes them sink?"

EXPERIMENT

- How can you make your boat move? Blow on it with a fan or through a straw.
- What makes your boat sink? Offer your child bowls full of pebbles, leaves, wooden blocks, and/or pom-poms. How many of these objects does it take for the boat to sink?

POUNDING NAILS

Before my children ever slammed a real nail into a real piece of wood, they practiced hammering golf tees into hard dirt with toy hammers. What a great way to build hand-eye coordination! My friend Liz, who happens to be a rock-star preschool teacher, taught me this nifty trick in her backyard, and I ordered my own big pack of golf tees as soon as I got home.

When I felt like my kids were ready, I taught them how to hold a real hammer and use it to drive nails safely through wood. A trick I picked up for keeping the nail steady is to hold it in place with a comb while hammering. This way, the child can easily hold the comb instead of the nail.

Some lumber yards and karate studios will give away small scraps of lumber, and it never hurts to ask. My best advice when you get started with woodworking is to keep it simple and start with one tool at a time: a screwdriver and a bag of screws, or a hammer and large, easy-to-hold nails. Make safety your first priority and go over the ground rules for using tools. Our rules: always wear safety goggles, and only use tools with adult supervision.

Supplies
- See individual invitations for specific items

INVITATIONS

Hammering Practice. Offer your child a small hammer and a box full of golf tees. Invite her to hammer tees into a piece of styrofoam, a pumpkin, a cardboard box, or a patch of dirt. Pound roofing nails into ceiling tiles. Hammer nails into a log. Draw a simple picture on soft wood and then hammer nails over the shape. Examples of soft woods include pine, spruce, balsam, cedar, fir, and redwood.

Attach Objects with Nails. Hammer metal bottle caps, bubble wrap, fabric, or precut craft wood shapes to soft wood with roofing nails.

Fastening with a Screwdriver. With the appropriate screwdrivers, fasten Phillips-head (crosshead) and flat screws into Styrofoam or soft wood that has prebored holes. Take an appliance apart (see page 124) and then screw the small pieces to wood.

Find a Real-World Project for Hammering. Hammer nails into a fence so you can hang your mud-kitchen utensils, hammer nails into a wall to hang pictures, and so on.

TAKE THINGS APART

The first rule of tinkering is to save all the parts.

—Paul Ehrlich, German scientist

What would a Tinkerlab book be without an experience in tinkering? A tinkerer is someone who experiments with materials and ideas to fully understand their capacities and then furthers her learning to find better solutions to current problems. At its core, tinkering is a hands-on way to explore how something is assembled in order to learn from or improve it.

Taking apart appliances, toys, and other familiar objects can help children understand how everyday products work, while also building comfort with tools like screwdrivers, hammers, and scissors.

You can tinker with just about anything, but stay away from appliances with glass or sharp corners. Also, steer clear of electronics such as microwaves, computer monitors, cell phones, and televisions, since many of them contain capacitors, which are objects that store electrical energy. If you accidentally hit one, you're in for a shock—literally. If you're not sure an item is safe, do a little bit of research first.

You may have some ready-to-donate objects around your home that are perfect for tinkering, or hit up your local second-hand shop for a treasure trove of possibilities.

Supplies: Favorite Objects to Tinker With
- Nonelectronic toys
- Household fan
- Mechanical clock
- Mechanical typewriter
- Rotary telephone
- Stuffed animals and action figures

Tools
- Flat-head and Phillips-head screwdrivers in multiple sizes (A lot of screws are tiny, so you'll want to have a miniature screwdriver set handy.)
- Miniplyers
- Tweezers
- Scissors
- Safety goggles
- Drawing paper and pencil to take notes or document the experience with drawings

INVITATION

Select your object and, if applicable, remove the batteries before you start. If you can find an assembly map or instruction booklet for the appliance online, print it out before you start and follow it as you take the object apart. Before you get started, put on safety goggles to protect your eyes from pieces that might fly around when they're disengaged. Set out the object along with your tools and invite your child to take the object apart.

As you disassemble the appliance, make educated guesses about the functions of each part.

EXPERIMENT

- Try to put the object back together. (This can be tricky, but not impossible.)
- After you take the object apart, use a low-heat glue gun to reattach the pieces in a new way.
- Hammer the individual pieces onto a wooden base (see page 122).

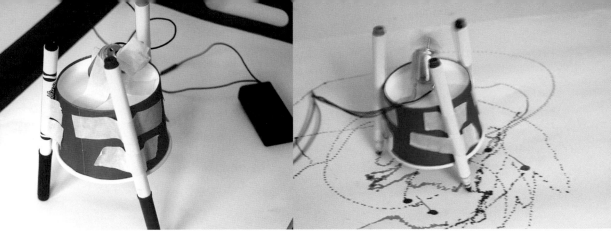

DRAWING MACHINE

Automate your drawings with your very own drawbot—a happy marriage of art and technology for little hands. If electronics take you outside your comfort zone, this simple project is an easy place to begin. I'm much more of an artist than an engineer, and if I can do this, you can too! The core of this simple robot is an offset motor, which is easy to find at any electronics store or online. Not only are these robots hilarious to watch, but the process of making them can empower children to see themselves as builders and engineers.

Supplies
- Plastic or paper cup
- Small offset motor (1.5–3 volts), also known as a toy motor
- Battery pack with wires and AA or AAA batteries
- Wire stripper (optional)
- Masking tape
- Electrical tape
- Weight, such as small change or a small screw. I like using dimes for this project.
- Three or four markers

INVITATION

The technical qualities of this project make it one for an adult and a child to build together. After you follow the steps for building your drawing machine, you can invite your child to change the colors or manipulate the motor and weight to change the direction and speed of the machine.

HOW TO BUILD A DRAWING MACHINE

1. Turn the cup upside down and securely tape three markers to it so that they form a tripod. This will elevate the cup off the table. Keep the pen caps on until your machine is ready to draw.

2. Put the batteries in the battery pack. Twist the wires of the motor to the wires of the battery pack by connecting the red wires to each other and the black wires to each other. It shouldn't matter if you connect matching wires. If your motor doesn't have wires, attach the battery pack wires to either of the motor's connectors. If there isn't enough exposed wire, use a wire stripper to reveal more. Fasten the wires together with a small piece of electrical tape.

3. With the motor and battery pack now connected, secure them to the top of the cup with masking tape. The placement of the battery pack and motor is a great area for experimentation.

4. Now you're ready to add a counterweight to the motor. This weight will help your Drawbot move. Notice that the motor has a thin cylinder of metal that protrudes from the top. This is the axis. Tape a dime or penny to the top of the axis. Without this added weight, the motor will simply turn like a fan.

5. Place your drawing machine on top of a paper-covered table.

6. Remove the pen caps and turn on your motor. Be amazed!

EXPERIMENT

- Try using different weights on the motor axis by removing or adding coins or other small objects.
- Discover new drawing patterns by adding more markers.
- Build the same type of bot on a different cup or object such as a Lego structure with wheels to see how this affects the look of the drawing.
- Move the motor to a new spot on the object and see if it changes the way the machine draws.

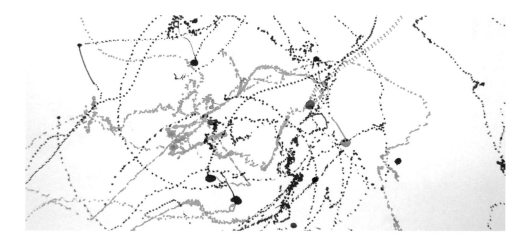

DIY ROBOT

Once you've made a drawing machine, you're ready to build something else with an offset motor. When I shared this challenge with my four-year-old, she decided she wanted to make a robot dog. This project follows the same principles used for making a drawbot, with the added bonus of encouraging a child to invent his own machine.

Supplies
- Small offset motor (1.5–3 volts)
- AA or AAA battery pack with wires
- Wire stripper (optional)
- Electrical tape
- Weight, such as small change or a small screw
- Selection of lightweight recyclable materials
- Markers and stickers to decorate
- Masking tape
- Clear tape
- Low-heat glue gun

INVITATION

Ask your child to brainstorm what else could be made with an offset motor. Some ideas: robot, train, car, animal. For the sake of this project, I'll refer to it as a robot, but feel free to encourage your child to create whatever he dreams up. When the motor is attached to an object, it can move. Based on what you witnessed with the drawing machine, talk about how you think this new object might move. Present your child with a selection of found materials for building a robot. Talk about how you can secure the pieces together with tape or a glue gun.

HOW TO BUILD A DIY ROBOT

Connect the motor to the battery pack. If the wires aren't exposed enough to allow this, use a wire stripper to reveal more of the wire, twist the wires of the motor and battery pack together, and secure them with electrical tape. Tape one or two dimes to the top of the motor axis to offset the balance of the motor. This will help the machine move. Without this added weight, the motor will simply turn like a fan.

Gather lightweight found materials to build your robot. Good choices include paper towel rolls, cardboard, paper, and parts of egg cartons. Fasten pieces together with tape or a glue gun.

Attach the motor to the robot with masking tape. Turn the motor on to test it. Does the robot move freely? Does it move forward or backward? If it's not moving at all, the materials may be too heavy. If it's going in the wrong direction, move the motor to a new spot or turn it around. Test and retest until it works the way you want it to. If other ideas emerge from this process, go ahead and build another robot or moving structure.

EXPERIMENT

You may want to add some light-emitting diodes (LEDs) to this project. LEDs are used commercially in remote controls, DVD players, and automotive dashboard panels; they're easy to find at electronics stores or online; and they're simple to work with. You'll need diffused LED lights of any color—5 or 10 millimeters—and CR2032 3V lithium batteries.

An LED has two poles on it: one is long and the other short. To make it light up, connect the long pole to the positive (+) side of a lithium battery and the short side to the negative (–) side. Keep the LED in place by wrapping a piece of electrical tape around the light and the battery. Secure the light to your robot. The LED will last one to two weeks.

DIY KIDS: BUILDING
TOMORROW'S INNOVATORS
THROUGH HANDS-ON MAKING

Grace Hawthorne, cofounder of *ReadyMade* magazine
and founder of Paper Punk

WHERE THE MAKER MOVEMENT STARTED

Sensing the void between *Martha Stewart Living* and *This Old House* for
our twenty- and thirtysomething friends, Shoshana Berger and I launched
ReadyMade in 2001. Right then, the resurgence of craft was percolating
underground; the DIY/makeover home show craze had not yet become a
pop culture phenomenon.

ReadyMade was one of the pioneers of the modern maker movement and
reuse design—when the act of making broke away from the traditional
home economics class. We had a pulse on the Gen XYers who were finding
it increasingly difficult to express themselves among the mass-produced
products in the marketplace. Because they were spending more time on
a computer at work, they ached to get away from the screen. The gratify-
ing feeling of using their hands to make something, personalizing it, and
calling it their own filled that void. Skills that were necessities for the
boomer generation were quickly becoming hipster hobbies.

MAKING IS A NECESSITY

The founders of *MAKE* magazine sought our counsel at *ReadyMade* before
they launched in 2005, focusing on the tech/science side of a male-
dominated maker culture. The publisher affectionately dubbed their
magazine "*ReadyMade*'s geeky cousin." Hands-on making has grown
exponentially over the past ten years since *ReadyMade* and *MAKE* came
on the scene. It is a definable cultural movement that will continue to

grow as our technological advances continue to create a digital divide between our authentic real-life experiences and the flat, two-dimensional, digital ones. Making reflects our need as a species to retain contact with nature, with ourselves, and with each other.

Regardless of any future technological innovation to come, our hands combined with our minds are the most potent, reliable tools for innovation in the century to come. The danger in becoming disconnected with the tangible world is that we lose touch with the virtues and characteristics that make us uniquely human, our sense of imagination and ability to make unique connections among disparate situations.

To think outside the box, you have to know what the box is. To change the world, we have to understand the world in which we live. Nothing replaces the authenticity and impact of an interaction or a learning experience in the real world.

MAKING AS A MIND-SET

The culture of making is more about a mind-set than a lifestyle. Makers have a curiosity for the built world around them. Understanding where things come from, how they are made, the potential of their function now and beyond, and what it takes to make something not only provides a context to appreciate the object at hand, but is essential to imagining infinite possibilities for the future.

To create maker kids, we have to be maker parents, maker people. That doesn't mean we have to be professional woodworkers. It spotlights our roles and influence as parents. Kids learn from watching us model behaviors. My entrepreneurial roots and maker genes were sown by my frugal parents, who were the first real environmentalists out of financial necessity. I grew up in a household that was about saving and stretching, not consumption and waste. My upbringing fostered a sense of resourcefulness and a deep appreciation for craftsmanship and quality. Although it may be more challenging to exemplify such virtues these days because we live in a disposable culture of convenience, tools for practicing maker behaviors are more readily available because of it.

THREE SIMPLE STEPS FOR EVERYONE

The good news is that having a maker mind-set is really simple and probably something you've already done or do regularly. Being mindful about creating maker opportunities, characteristics, and biases is possible with a quick glance around at everyday things. Here's an example:

- Look around you. It's as simple as taking the time to look at a cereal box before banishing it to the recycling bin. That simple cereal box is full of learning opportunities.
- Pique your child's curiosity with prompts. Some simple starter questions:

What is this made of?
How was this made?
Where was it made?
Can you make this yourself?
Where else have you seen this material?
How many different ways could we reuse this box instead of
     discarding it?
What other ways could we package/store cereal?

- Turn curiosity into action. Some simple activities to do with the box:

Explore the materiality of the box—tear it with your hands, cut it
     with scissors, fold it, and so on.
Make it move—cut a pinwheel or flying disc out of it.
Make it hold something—fashion a mail holder, hamster maze, or
     gift box.
Make believe with it—use it as a telescope, treasure map, or tele-
     phone.
Make it share—turn the graphics from the box into a cool greeting
     card.

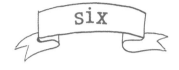

CONCOCT

When my oldest daughter turned three, I noticed how much she enjoyed mixing concoctions, so we spent the better part of our summer building and playing in our outdoor mudpie kitchen. I found a shady spot near a water source and plenty of dirt, and she was in concoction heaven. Cautious of investing too many resources in something my child might never use, I built a very simple version of a garden kitchen for us to test. It consisted of two apple crates, a few bowls, and some utensils. This little kid-driven outdoor space became concoction command central.

As children collect and mix ingredients and marvel at the surprises that emerge from their mixtures, they become small alchemists, bakers, scientists, and inventors. The processes of observing and formulating hypotheses are at the heart of these experiments, and children can get lost for hours with their arms covered in magic potions of flour, water, and vinegar.

These experiments could mark the career beginnings of future scientists or chefs, but perhaps more important, they show children how to ask questions, test ideas, and think flexibly. For example, when a child asks, "What will happen if I mix dirt with water?" he begins to think more deeply about the world of cause and effect. He'll exercise his flexible thinking skills when he has the freedom and time to explore how ingredients work together through improvisational trial and error.

This last idea—improvisation, or having the ability to think quickly and act on the spot—is at the heart of the experiments in this section. When children are flexible in their thinking, they're better equipped to navigate and respond to the rapid changes that are sure to happen in the world around them.

YES, AND... HOW TO IMPROVISE WITH CHILDREN

An interview with Dan Klein,
improviser, Stanford University

Q: :Can you give us a little primer in the basics of improvisation?

A: Improvisational theater is an art form that emerged as a way to make up stories, collaboratively, in front of an audience. In order to do that, improvisers train to accept all offers. There is no time to sort and judge and evaluate—improvisers just say "yes" and then justify it afterward. Improvisers practice being present, observant, and delighted by the world. They try to say "what if" instead of "we can't."

Q: How can families use the tenets of improvisational theater in their daily lives?

A: In the family ecosystem, there are two types of improvisers: child improvisers and parent improvisers. Child improvisers are creative geniuses. They are able to imagine and play in a world of make-believe that parents can only long for. While parent improvisers can tap into this world, they are equally saddled by judgment, distractions, and experience. They are burdened by interpretation and meaning. But that's okay, because parents also have wisdom and big-picture awareness. And, of course, they were once child improvisers themselves.

When the child improviser throws out an offer, such as "Let's play camping!" the job of the parent improviser is to say, "Yes, and let's set up a tent"—to join the child as a collaborator in the world he or she sees. It's not the parent improviser's job to be creative but to give just enough to inspire the child improviser. Sometimes I lead creativity workshops with kids and adults together. We play a game called Yes, Let's! One person suggests a pretend activity: "Let's eat ice cream!" Everyone enthusiastically says, "Yes, let's!" and pretends to eat ice cream, until someone makes the next suggestion. I am always struck by how differently the kids and adults say "yes." When I say, "Let's all dig for buried treasure," the adults look around for a shovel, consult the map, check to see who is watching. The kids dive to the ground and dig furiously with both hands.

Q: What are some of your other favorite improvisational games to play with children?

A: Giving presents. Begin by saying, "I have a pretend present for you," and hand the child an invisible box. He might open it and name what he sees. To whatever the child says is in the box, say "yes" and add to it. Tell him where you got it, or point out a feature or detail. If he seems reluctant to open the box, you may need to give him some clues as to what it could be. Your job is to give just enough to inspire him. You could do this with a comment such as "You know how you always wanted another pet," or "I found this at the magic store."

Word-at-a-time. Make up a story by beginning with one word and then inviting your child to add a word. Go back and forth (or in a circle around the dinner table) until you agree that the story is complete. Depending on the age of your child, she might get attached to an idea of where the story is going. That's okay. Your job is to say "yes" and add the next word.

Remember that time we. One of you starts by saying, "Remember that time we swam to the bottom of the ocean (or 'got lost at the zoo' or 'stood in the rain' or whatever)?" Another family improviser says, "Oh yeah, and we found a sunken ship (or 'wandered into the elephant pen' or 'got soaked')!" Then you just go back and forth, saying "Oh yeah, and . . ." to add to the story.

POTION STATION

The meeting of two personalities is like the contact of two chemical substances: if there is any reaction, both are transformed.

—Carl G. Jung, Swiss psychologist

Most kids are mesmerized by the process of pouring vinegar onto baking soda, watching it fizzle, and then sprinkling more baking soda on top to start all over again. When my two-year-old and I first tried this out and she plowed through all of our cider vinegar, I felt generous and handed over bottles of wine and balsamic vinegar so we could compare the results. The next week I made a point of adding a gallon of vinegar and a few boxes of baking soda to my shopping list, because it was obvious that this would be a staple of our creative diet for a while. Since that first introduction to vinegar and baking soda, known to many as the Volcano Experiment, we've re-created this and other similar mad scientist experiences more times than I can remember. There are a lot of ways to make potions, and this experiment includes a few of our favorite ingredient combinations.

- White vinegar (although any kind will work)
- Baking soda
- Bowls or small jars
- See individual invitations for specific items

INVITATIONS

Cups and Bowls. Place a few bowls in a sensory tub and scoop a teaspoon of baking soda into each bowl. Fill a cup or small pitcher halfway with vinegar. Invite your child to pour the vinegar into the bowls and see what happens. Once the fizzing settles down you can simply add more baking soda to each bowl and more vinegar to the pitcher, then start all over again.

Colorful Fizzing Tray. Fill a baking dish with baking soda. Fill small bowls with vinegar that's tinted with food coloring. Offer your child an eyedropper to use in adding colorful vinegar to the tray.

Spray Bottle. Sprinkle baking soda across the bottom of a sensory tub. Pour vinegar tinted with food coloring into a small spray bottle that your child can control easily. Invite her to spray the vinegar all over the baking soda. For another twist, add droplets of food coloring to the baking soda before spraying.

EXPERIMENTS

- Stir dish soap and food coloring into the vinegar before adding it to the baking soda for some extra bubbly action.
- Add glitter too. Why not?
- Replace the baking soda with baking powder. How are the results different? Or are they the same?
- Replace the vinegar with another acid like lemon or lime juice. How is the reaction different?
- Drop a piece of hard candy into the potion. What happens to it?

DOES YOUR CHILD HAVE A FAVORITE WAY TO MAKE CONCOCTIONS?

Baking soda and vinegar is always a winner with my children. We've tried it in many different ways (of course, that might be because Mommy likes it so much).

—Jillian R.

My children love to make a potion lab, mixing up ingredients from the kitchen into soup and medicine. When they were younger, they enjoyed the sensory nature of the materials and all the mixing. Now that they are older, they always involve role-play as they experiment—sometimes working in an orphanage and feeding all the children with their potions, sometimes being magicians making medicine.

—Cathy J.

Lucy is obsessed with mixing colors. Any artwork she's doing, especially with paints, will eventually turn into a color-mixing experiment! It's also fun watching her try to mix crayons, colored pencils, oil pastels, and so on and wonder why they don't mix as well. This leads us to lots of fun conversations and theories. I've shown her the reverse as well, separating the colors. We use coffee filters and water to separate the pigments in markers. She can see right in front of her what colors were used to make the marker she's drawing with.

—Chelsey M.

My children like mixing all the colors up and then sticking them outside. They also love freezing and defrosting water in a cup—sometimes with toys trapped in the ice—over and over again. I think they like these activities because they are "experiments" that they can do on their own, and I allow them to get on with it.

— Maggy W.

The kids make the best concoctions, mainly outside, because I like to send them out when I'm cooking. The backyard belongs to the kids, so they are free to do as they wish. There is access to water, plenty of weeds, a sandbox, tree bits, bugs, and numerous containers. A wizard's dream.

—Danielle A.

My boys have "potion kits." Each of them has a plastic shoebox with an assortment of old hotel shampoos, empty spray bottles, droppers . . . anything nontoxic they can make a potion with. The rule is that potions don't go *on* or *in* anybody. I have them keep the boxes outside, and they work for hours creating things like stink potions.

Alissa M.

My son is a master of concoctions. He loves to mix everything he can find in the kitchen, mostly water, oil, salt, sugar, spices, cocoa, and coffee, but also wood pieces, branches, leaves, stones, seeds, soap, and toothpaste. He loves to wait for days and see what happens (if mildew appears, he is even happier). My kids learn about miscible and immiscible things, but they also learn to wait patiently for days, developing hypotheses as the days pass and events occur, accepting that things do not always go as they expect, tolerating frustration, trying to remember and re-create formulas, repeating processes. (Isn't this the scientific method?)

—Angela A.

GOOP

Oh goop, what messy and mesmerizing fun you are. Goop has the strangest consistency—it's liquid one moment and solid the next. Children love it, and adults who have never experienced it will marvel at its texture. You've been warned!

Supplies
- 12 to 16 ounces cornstarch
- 1 to 1½ cups water (start with 1 cup and add as you go if needed)
- Sensory tub
- Toys for playing, scooping, and filling, such as spoons and little bowls
- Food coloring (optional)
- Big storage container or water table

INVITATION

To make good goop, you need a ratio of two parts cornstarch to one part water. Rather than present my girls with goop that's ready to go, I like to involve them in the process of making it. It's not only fun, but it helps them learn about the separate properties of cornstarch and water and what happens when they're combined.

Start by pouring the cornstarch into the middle of the tub. Feel it, squeeze it, sift it through your fingers, and talk about the texture. Next, pour in half the water and mix it with a spoon or your hand. How does it feel? Add the rest of the water and continue mixing. Again, talk about the texture.

Once this exploration has run its course, introduce a couple spoons or mixing tools to scoop, stir, and mix the goop. If you're feeling brave, invite your child to walk barefoot through the goop. As a last step, invite your child to squeeze in a couple drops of food coloring. Stir it in, spin it around in swirls, and enjoy your colorful goop.

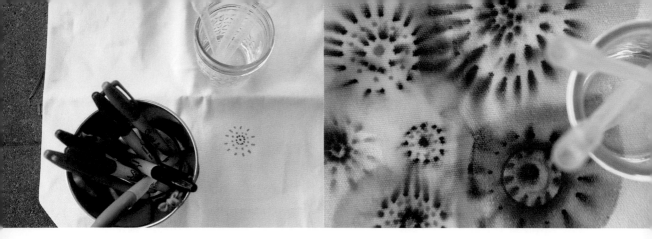

MARKER EXPLOSIONS

These stunning firework-like explosions are made by dissolving permanent markers with small quantities of rubbing alcohol. But how does it work?

Washable markers are water-based, which means that they will dissolve in water. Nontoxic permanent markers, on the other hand, are not water-based and water can't dissolve them. This is why it's near impossible to wash permanent marker from clothes. These indelible markers are a composition of pigment (the color), solvent (a nontoxic alcohol that dissolves the pigment), and resin (a glue-like polymer that helps the ink stick). Because the resulting compound is alcohol-based, it can only be dissolved with a similar compound, such as alcohol or acetone.

Supplies

- Piece of light-colored cotton fabric, such as a plain T-shirt, cut-up sheet, or canvas bag
- Piece of cardboard, cut to fit beneath the fabric
- Permanent markers.
- Small jar filled with rubbing (isopropyl) alcohol
- Small jar filled with water
- Eyedroppers

Note: Not all permanent markers are the same. To keep your family safe, look for products that ACMI certifies with its AP or CL seals. Sharpie is a popular brand in my house.

INVITATION

Place cardboard beneath the piece of fabric. If necessary, clip it in place to keep it from moving around. Next, draw on the fabric with permanent markers. Concentric circles of dots will make for a firework explosion, but try other designs as well. Squeeze rubbing alcohol into the eyedropper and release it onto the drawing. If you drop it in the middle of the circular design, it will give the appearance of a firework. Try other designs and see what results you get.

EXPERIMENT

Water, Markers, and Coffee Filters. Draw on coffee filters with washable markers. Fill an eyedropper with water and squeeze drops in the middle of the drawing.

Are Black Markers Really Black? Draw on a coffee filter with a black washable marker. Drop water on the marker and discuss the colors you see.

PERMANENT-MARKER CLEANING TIP

Depending on the stained surface, you'll need a nonpolar solvent such as acetone-free nail polish remover, isopropyl alcohol, or Amodex stain remover to remove the trouble spot.

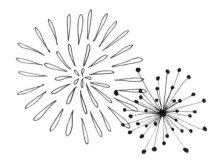

MAKE YOUR OWN (SEMIEDIBLE) PAINT

Is it possible to make your own paints? Experiment and find out! The process of making paint empowers children to take matters into their own hands if or when resources run dry. I'm a fan of store-bought paints, but knowing every single ingredient that's included in our homemade paints give me peace of mind when I see little hands covered with it. The following recipes use household staples and are very easy to whip up. See Paint Experiments (page 81) for ideas on what you can do with your homemade paint.

POWDERED MILK PAINT

This paint will have a yogurt-like consistency and a matte finish when dry.

Supplies
- 2 tablespoons nonfat dry (powdered) milk
- 1 tablespoon water

- Small mixing bowls
- Food coloring or natural dye
- Spoons for mixing

For each color, mix the powdered milk with the water until it's thick like yogurt. Add food coloring or natural dye to get colors you like. Add more or less water to reach the desired consistency. This paint can be covered and stored for up to four days in the refrigerator.

SWEETENED CONDENSED MILK PAINT

This paint goes on thick and dries shiny.

Supplies
- Sweetened condensed milk
- Food coloring
- Small mixing bowls
- Spoons for mixing

Pour a couple tablespoons of sweetened condensed milk into a bowl. Add a few drops of food coloring and mix with a spoon. This paint is edible, goopy, and delicious. To address the elephant in the room, I let my kids lick their spoons before we paint, then they get one more taste when we're done. This paint goes on thick, so have a damp towel nearby to wipe sticky hands. When the paint dries it will be shiny (and quite dry—not sticky). These paintings will hold up beautifully over time. Store extra paint in the refrigerator for up to five days.

EGG TEMPERA PAINT

This paint will have a subtle sheen when dry.

Supplies
- Eggs
- Small mixing bowls
- Bowl to hold egg whites
- Spoons for mixing
- Food coloring or liquid watercolors

Use one egg yolk per color. Separate the yolks from the whites. Drop a yolk into a small bowl, add a few drops of food coloring or watercolor, and mix. Over time, egg tempera may crack, but it's a fairly stable paint. Egg tempera paint does not store well and should be used the same day it's made.

SQUEEZABLE FLOUR-AND-SALT PAINT

This thick paint dries to a raised, puffy texture.

Supplies

- ½ cup flour
- ½ cup salt
- Mixing bowl and spoon
- ½ cup water
- Three plastic bags
- Food coloring
- Scissors
- Squeeze bottle (optional)

Stir the flour and salt together in a bowl. Add the water and mix until smooth. Divide the mixture into three plastic bags. Add food coloring to each until you reach the desired color. Seal the bags and squish the paint around until it's well mixed. Snip off a corner of each bag and squeeze the paint directly onto paper, or transfer the mixture to a squeezable bottle. This paint shows up well on dark paper. Extra paint can be covered and stored at room temperature. The salt should act as a preservative, but you'll want to throw away any paint that grows mold.

FLOUR PAINT FOR FINGER PAINTING

This thin paint dries to a matte finish.

Supplies

- ½ cup flour
- 1 tablespoon salt
- Mixing spoon
- Small pot
- ¾ cup cold water
- ⅔ cup hot water
- Food coloring or liquid watercolors

Combine the flour and salt in the pot. Add the cold water and mix until smooth. Add the hot water and boil until the mixture is thick and smooth like paint.

SIDEWALK PAINT

This thin paint looks bright and chalky when it's dry.

Supplies
- 1 cup cornstarch
- 1 cup water
- Mixing bowl
- Wire whisk
- Muffin tin or small containers
- Food coloring or liquid watercolors

Whisk the cornstarch and water together until smooth. Pour this mixture evenly into a muffin tin or small containers. Mix food coloring or paint into each cup. Paint directly on concrete with a paintbrush. The paint brightens as it dries and will wash away with water. We haven't had it stain our sidewalks, but proceed with caution if you're using it on a precious surface. Store covered paint at room temperature. Revive dried paint with more water.

Renaissance artists made their own paints in studios that probably felt a lot like chemistry labs. The paints they used were made from two compounds: binder and pigment. They often worked with egg tempera paint, similar to the recipe shared here, which was made from a binder of egg yolks and a pigment of minerals and colorings such as saffron (yellow), clay (brown or red), or finely ground rocks such as lapis lazuli (blue).

SLIME

Slime is squishy and gooey, and although it magically molds to whatever tries to contain it, it doesn't stick to skin. It's so economical and easy to make that I marvel that it took me so long to discover it. The big challenge lies in finding borax (sodium borate), but once that's secured, you'll undoubtedly find lots of excuses to make big batches of silly slime.

Borax is a powdered household cleaner and laundry booster that you can usually find in the detergent section of the store. It's made from a borate mineral that is found in evaporated playa lakes in arid regions. Please use your best judgment when using this material with small children, as borax is not edible.

The recipe is a combination of white glue, water, and borax. Here's how it works: When you combine common white glue (polyvinyl acetate, or PVA) with a borax-and-water solution, the borax binds the glue molecules so they form larger molecules called polymers. Squishy, bouncy slime results from this new compound's ability to absorb a large quantity of water. The texture is irresistible and inspires many ideas for exploration. A small caution: Slime loves to bond with fabric and rugs, so be sure to play with it over noncarpeted floors and away from upholstered furniture.

Supplies

- ½ cup white glue
- ½ cup water
- 1 teaspoon borax mixed into ½ cup water
- Spoon
- Medium-sized mixing bowl

INVITATION

Invite your child to help you stir the glue and water together in the mixing bowl. Slowly add a small amount of the borax-water solution to the bowl and mix. Add more of the borax solution until the slime pulls away from the bowl and comes together. You will most likely not use the entire solution.

Play with your slime!

EXPERIMENTS

- Cut the slime with child-safe scissors.
- Elevate a colander on some blocks or books. Place the slime in the colander and watch it ooze through the holes. This activity is entirely satisfying and good for building patience.

- Fill a muffin tin with slime. Remove your "muffins" and see how long it takes them to lose their shape.
- Challenge your child to invent things to do with it.

MAKE YOUR OWN BOUNCY BALL

Ask your child to pour ⅛ cup of glue into a small mixing bowl. Add 1 teaspoon of water and a few drops of food coloring, and mix well. With a spoon, add a small amount (roughly ¼ teaspoon) of the same borax-water solution from the experiment to the glue solution and stir. The solution should be getting sticky. Keep mixing in small amounts of the borax solution until the whole mixture joins together and loses its stickiness. It should feel like silly putty when it's done. Roll the thick slime into a ball and play with your newly formed bouncing ball.

ICE AND SALT EXPLORATION

In the philosophic sense, observation shows and experiment teaches.

—Claude Bernard, French physiologist and science historian

When you live in a warm place like we do, ice experiments never grow old. On the other hand, there's nothing like a good, snowy winter for making ice experiments come to life. This experiment teaches children about the physical relationship between salt and ice—salt lowers the freezing point of ice and will cause it to melt at contact. The cause and effect of this experiment also encourages children to ask questions and test various combinations of salt, water, color, and ice. Consider this a launching pad for other experiments. After running through it, my four-year-old wanted to test the reaction of sugar with ice and then created a concoction of salt and watercolors for a painting experience.

Supplies

- Water
- Glass or plastic containers for freezing ice
- Rock salt or kosher salt
- Food coloring or liquid watercolors
- Eyedropper
- Large tray with sides

PREP

Fill a few milk cartons, plastic containers, or freezer-safe bowls about three-quarters full with water. Place them in the freezer—or outdoors if it's below 32°F (0°C)—overnight until they contain solid blocks of ice.

INVITATION

Release the ice from the containers and place it in the tray. Pour liquid water-colors into small bowls or the compartments of an ice cube tray, and fill a small bowl with rock salt.

Offer your child the tray with ice and the bowl of salt. Invite him to sprinkle the salt over the ice and talk about the chemical reaction that occurs (the salt will bore holes in the ice). Next, offer your child the watercolors and an eyedropper, and invite him to squeeze watercolors into the cracks in the ice. The colorful results are spectacular. Encourage color mixing, the addition of more salt, and other experiments.

EXPERIMENTS

- Try adding other ingredients such as sugar, flour, or hot water to the ice. Is there a reaction?
- Set up a few different trays of ice, each one with a different type of salt such as rock salt, table salt, and sea salt. How do each of them interact with the ice?

ICE CREAM IN A JAR: AN EDIBLE INVESTIGATION

While ice cream is readily available, making your own can teach children about how it is made and thrill them with the fruits of their own labor. Your child may have more tenacity for this than mine, but I warn you that it requires a fair amount of strength, and you may end up doing most of the work.

Ice begins to melt at 32°F (0°C). Salt lowers that melting point and makes the ice even colder than it was in your freezer. This helps the ice cream freeze when it's surrounded by the ice. Without the salt, the milk and sugar would just get very cold. You could also make ice cream with table salt, but the larger crystals of the rock salt take longer to dissolve in the ice and help the cream come together more evenly.

Supplies (makes one scoop)

- ½ cup half-and-half *or* ¼ cup heavy whipping cream and ¼ cup whole milk
- ½ teaspoon vanilla
- 1 tablespoon sugar
- Small canning jar with tight-fitting lid
- Gallon-size freezer bag with a tight seal
- ⅓ cup kosher or rock/ice cream salt
- Hand towel
- Approximately 4 cups ice

INVITATION

Ask your child if he would like to make his own ice cream. Chances are he'll say, "Yes!" Invite him to stir the first three ingredients inside the canning jar. Once they're well mixed, seal it up. Place the canning jar in the freezer bag and fill the bag with ice. Then pour the salt over the ice. Squeeze as much air from the bag as you can and seal it tightly.

Place the bag in the center of a hand towel and cinch the towel closed at the top. Holding the top of the towel, shake it for about eight to ten minutes, or until ice cream forms. I'm never convinced that it should take that long and invariably check around the halfway mark, but I've never been blessed with five-minute ice cream.

Discuss the ice cream–making process with your child:

- How does the mixture look after five minutes?
- How long did it take the ingredients to come together into ice cream?
- How does it taste?
- Does it taste different from ice cream you usually eat? How?
- What could you add to the ice cream to change the flavor?

EXPERIMENTS

- After the ice cream is ready, add some mix-in flavors like baking cocoa, coconut flakes, peanut butter, or chocolate chips.
- Would this work with full-fat yogurt instead of cream or half-and-half? Try full-fat yogurt alone, as well as a combination of heavy cream and yogurt.

FROZEN CARBON DIOXIDE

Dry ice, also known as frozen carbon dioxide, dissipates into carbon dioxide gas when it melts. It's called "dry ice" because it turns directly into a gas from its solid state without ever becoming a liquid. Thus, there's no puddle of water when it melts. This process is called *sublimation*.

You can often find dry ice at grocery or hardware stores and sometimes at ice cream shops. The temperature of dry ice is –110°F (–78°C), so be careful when using it. It will burn your hand if you touch it. But if you take precautions and your child is prepared to pay attention to safety rules, dry ice is a cool material to experiment with.

Supplies
- Four pieces dry ice, each the size of a button mushroom
- Metal thermos
- Warm water
- Kitchen tongs
- Large tray with edges

PREP

Go to the store with a cooler, or call ahead to see if the store has a safe way for you to get the ice home. If your dry ice comes in a large chunk, ask the shop to break it into smaller pieces for you. You don't want to break the ice with a hammer or your hands.

Be safe and follow all the directions that come with your dry ice. Once home, you'll want to have a thermos and tongs ready to conduct this experiment safely. Do *not* touch the dry ice with your bare hands, and *always* use dry ice outside or in a well-ventilated area to avoid asphyxiation.

INVITATION

Place a thermos in the middle of a large tray with edges, then fill the thermos with about a cup of warm water.

Remind your child about the dangers of touching dry ice. Pick up a small piece of ice with your tongs and ask your child some questions about the experience so far:

- "Do you hear that sound? What does it remind you of?" (If you hear a loud humming noise, the pressure of the ice is making the tongs vibrate.)
- "What do you think will happen when I drop this dry ice into the warm water?"

Drop the ice into the thermos with the tongs. When the ice hits the water it will begin to smoke. The smoke is okay to touch. Ask your child, "What happened when the ice touched the water?"

EXPERIMENTS

- Blow on the smoke.
- Squeeze some dish soap into the thermos. What happens?
- Add some drops of food coloring or liquid watercolors on top of the dish soap mixture.
- Empty the thermos and fill it with about a cup of milk. Drop a piece of dry ice in the milk and compare the results with those from your water experience.
- Try the experiment again with warm water in one thermos and cold water in another. Ask your child, "How does the dry ice react to differences in temperature?"

YEAST AND SUGAR
EXPANSION

One fine bread-making afternoon, my preschooler asked me about yeast, so we got into a discussion about how it helps the bread rise. She wanted to know more, which gave birth to this experiment. Yeast is inactive in its dry state. The addition of warm water activates the yeast, and after a few minutes in warm water, you'll see bubbles rise to the surface of the solution. Yeast feeds on sugar, and in bread making it will convert the starch in flour to sugar. Adding a small amount of sugar to the yeast–warm water mixture will speed up the fermentation process. Try including your child in a bit of bread making, and conduct this experiment as a way to support real-world learning.

Supplies
- Standard size balloon (roughly 9" to 12")
- Narrow-neck bottle that you can fit the balloon over
- 2¼ tablespoons active dry yeast
- 2 tablespoons sugar
- 1 cup warm water

- Mixing bowl and funnel (or a cocktail shaker, which is our preference)
- String to measure the balloon (optional)

INVITATION

Set your bottle up inside of a shallow storage container, with the balloon nearby. Invite your child to help you stir the yeast, sugar, and warm water in a mixing bowl. Smell the concoction and ask your child what it smells like. My kids said "poop," so consider yourself warned.

Once the mixture dissolves, pour the yeast mixture into the bottle using the funnel, and cover the bottle with the balloon. You may see the concoction bubbling, which means it's making carbon dioxide. Wait for the balloon to rise. This may happen quickly, or it may take a little while. Not to worry, the balloon (probably) won't pop off and explode, but it could be messy when and if you remove it from the bottle.

If you like, use a string measure the girth of the balloon as it grows. Place the bottle in the sink and remove the balloon. Watch the bubbles slowly pour over the top—big fun if you're a kid!

EXPERIMENTS

- Try the same thing with food coloring added to the water.
- Set up more than one bottle and test different ratios of water to sugar to yeast.
- Compare the results with different water temperatures. Set up multiple bottles at the same time and see which balloon grows the fastest.
- Replace the sugar with sugary soda, maple syrup, corn syrup, or juice.

NAKED EGG EXPERIMENT

Nature is a source of truth. Experience does not ever err, it is only your judgment that errs in promising itself results which are not caused by your experiments.

—Leonardo da Vinci, Italian artist, architect, and engineer

A naked egg is simply an egg without a shell. How does this happen? A good soak in vinegar will remove the shell, while the egg itself is held together by a thin membrane that the vinegar cannot break down. The shell is made of calcium carbonate, a base that dissolves when it mixes with an acid—in this case, vinegar. You'll see the same principle in action when mixing baking soda (base) with vinegar (acid).

The process of stripping an egg from its shell is fascinating, and holding a slippery, naked egg is an experience in itself. My daughter and her friend sat down one summer afternoon to set this experiment up, and the two of them were very curious to see what would happen after their eggs sat in vinegar for twenty-four hours. As soon as the whole experiment was done, the two of them wanted to do it all over again. So we did.

Supplies
- Raw egg
- Clear vinegar (I use cider vinegar because it's clear and inexpensive.)
- Glass jar that will hold the egg (A clean pickle jar is great for this job.)
- Food coloring or liquid watercolors (optional)
- Tray or plate

INVITATION

Set up your supplies and tell your child you're going to make a naked egg. Invite her to place an egg carefully in the jar. If the opening isn't large enough to allow a hand, use a spoon or something similar to slide the egg in. Pour vinegar on top of the egg to cover it. Add a few drops of food coloring or watercolor until you reach the desired color (optional). Leave the egg in the solution for one full day, rotating it occasionally so that all sides get submerged in the vinegar.

While you're waiting for the eggshell to dissolve, discuss the process with your child. Here are some of my favorite guiding questions:

- After covering the egg with vinegar—"What do you think will happen to the egg after it sits in vinegar overnight?"
- The next day—"What do you notice about the egg today? How does it look different from yesterday?"
- After carefully removing the egg from the jar and inviting your child to touch it—"How does it feel? What do you think happened to the shell?"

At this point, the egg can be *carefully* bounced on the plate. Whoa, a bouncing raw egg! It's held together by the membrane, so be ready to clean up a mess if it gets bounced too hard.

EXPERIMENT

Try this experience with three different solutions: vinegar, corn syrup, and salt-water (or anything else you can dream up). What happens?

HOMEMADE BUTTER: AN EDIBLE INVESTIGATION

If you shake a sealed container of heavy cream long enough, the drops of fat that are usually suspended in the liquid smack against each other, eventually stick together, and then miraculously form butter. Not only that, but if you use good-quality cream, it may be the most delicious butter you'll ever taste. This edible concoction gets high marks for wowing kids with its magic, while introducing them to the possibility of making their own food from scratch.

Supplies
- Glass jar with a tight-fitting lid (I like canning jars for this project.)
- Heavy whipping cream
- Marble (optional)

INVITATION

Ask your child if she would like to help you make butter from scratch. Start with a clean glass jar. Pour the cream into the jar, filling it about a quarter of the way, which will allow room for shaking. Add a clean marble if you have one; this will speed up the process, but it's not critical to getting the job done.

Shake continuously until the cream divides into butter and "buttermilk." This can take anywhere from five to ten minutes. You'll know the butter is ready when it comes together and clunks in one piece around the jar. It can take some time; just be patient.

Scoop out and pat the butter into a bowl or mold. Save the sweet buttermilk for other recipes.

LEMON INVISIBLE INK

Invisible ink always makes me think about passed notes, childhood forts, summer camp, and spies. Not that I ever made invisible ink in forts, camp, or spy school, but passing secret notes between my grade-school friends holds a sweet spot in my heart. My little ones are too young to pass secret notes but not too young to enjoy the goodness of making invisible pictures come to life through the magic of heat. Citric acid weakens paper, and when this acid-soaked paper is heated, the acid burns and turns brown before the paper does. This experience of painting with an acid and bringing it to visual life with heat is a rich combination of art and science.

Supplies
- Lemons
- Two small bowls
- Juice squeezer or fork to juice your citrus
- Cotton swab and/or small paintbrush
- Card stock: white and colored (optional)
- Heat source, such as an iron or a lightbulb

PREP

Place a bowl of squeezed lemon juice alongside brushes (or cotton swabs) and paper.

INVITATION

Invite your child to paint on different papers with the lemon juice.

Heat the paper with an iron, a hair dryer on the highest setting, a lightbulb, or another heat source. Be careful that you don't hold it there too long or the paper could catch fire. We've tested all of these heat sources, and the iron worked best for us.

EXPERIMENT

Try this same process with other liquids such as milk, orange juice, white wine, and apple juice. Make a chart to track your results. How do these other ingredients dry on the paper and react to the heat source?

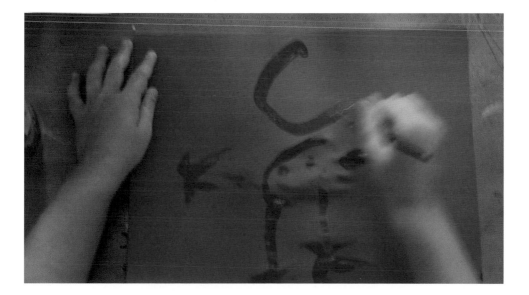

GLITTERY EGG GEODES

Two sodium atoms are walking along the street when one stops and says, "Oh my goodness, I think I've lost an electron!" "Are you sure?" asks his companion. "Yes," replies the first sodium atom. "I'm positive."

—Anonymous

These gorgeous, glistening egg crystals are about as much fun for parents to make and admire as they are for kids. When I first spotted these on the ever-so-inspiring Martha Stewart's blog, I knew our science-meets-art family would have fun testing various solutions to create our own crystals. Through a process of diffusion, salts pass through the permeable eggshell, but what will the borax do? Work your way through this experiment to find out. I'll share three solutions (to fill six eggs) that will make different crystals to chart and compare, but I encourage you to experiment with other ingredients too.

Supplies
- Six or more eggs
- ¼ cup kosher salt
- ½ cup sea salt
- Borax

- Other substances that can be tested for crystallization, such as sugar, epsom salts, cream of tartar, or baking soda
- Egg carton or minimuffin pan
- Food coloring
- 1½ cups hot water
- Spoon
- Cup or bowl for mixing
- Funnel (optional)

PREP

Tap a knife around the eggs to break the tops off. Empty the egg contents into a bowl (save them for breakfast!), and clean the empty shells with water. Gently run your finger around the inside of the eggs to remove the membrane, as it can discolor the crystals as they form. Place the clean, empty eggshells in the muffin tin or nonporous egg carton to dry, and move on to the alchemy part of the project.

INVITATIONS

Invite your child to help you mix up three solutions. In one jar, make a kosher salt solution by mixing ½ cup of hot water with ¼ cup of kosher salt until the salt more or less dissolves. There will still be salt at the bottom of your cup, but that's okay.

In another jar, make a sea salt solution by mixing ½ cup of hot water with ¼ cup of sea salt until the salt dissolves. Add more salt, bit by bit, until the water is saturated. You'll know you've hit the right point when no more salt can dissolve.

In a third jar, make a borax solution by mixing ½ cup of hot water with ¼ cup of borax.

Keep track of your solutions by making a handy chart. Add a couple drops of food coloring to each solution, then color code your chart to correspond with the solutions.

Place eggshells in an egg carton or minimuffin pan. If the egg carton is porous some of the liquid could leak through. Place it on a baking try that can contain the mess.

Pour the liquid into the empty eggshells—each solution will fill two shells. Wait one to five days for the solutions to evaporate and the crystals to form fully. This requires some patience, which makes it a good time to introduce your child to documenting experiments. Visit your geodes every day and mark your findings on your chart. Compare how the different solutions react with the eggshell.

NATURAL DYES

While my three-year-old and I were cooking down onion skins to make yellow dye one day, I asked her, "If we wanted to make yellow dye, what other ingredients could we use?" She had fun considering all the yellow things we had in our kitchen and garden and came up with suggestions such as flowers, banana skins, and lemons. While not just any yellow thing will make yellow dye, this early foray into hypothesizing showed her understanding of the challenge at hand.

It's important to state up front that this will undoubtedly be a parent-led project. Kids get excited about the first steps but tend to lose steam during the cooking process. But keep these recipes handy since they're useful for many of the other projects in this book.

Dye making is not an exact science, and what works in my kitchen may not work in yours since the intensity of color in vegetables and spices varies greatly. However, making dyes from natural materials helps children understand that they can create their own art supplies without a trip to the store, gives them a sense of history as they consider how artisans once had to create their own coloring agents, and encourages them to set up their own kitchen experiments as they explore a world of color through extracted pigments.

Once the dyes are finished, you can use them to color eggs, paint on paper, or color playdough. They will oxidize over time, which means they'll change color or fade. Professional dyers often add a *mordant,* a substance that helps set or alter the color of the dye at some point in the process. Examples of mordants are alum, tannic acid, acetic acid, and sodium chloride; white vinegar and citric acid also work and are what I suggest using for these dyes.

Supplies
- Vegetables, spices, flowers, and plants to experiment with
- Mordant, such as white vinegar or citric acid (lemon juice, for example)
- Big pot
- Cooking spoon
- Water
- Strainer
- Jars to store dyes

INVITATION

Talk with your child about making natural dyes and ask her to guess what you could use to make various dyes. Consider what color dye you would like to make. Refer to the recipes that follow for ingredients and quantities. Place the ingredients in a pot and cook over low heat for about thirty minutes. Strain the liquid into a bowl and cool before using. Save extra dye in sealed jars in the refrigerator for up to one week.

Natural Dye Recipes

Light green: ½ cup fresh rosemary + 2 cups water; skin from 3 red onions + 4 cups water

Orange/yellow: Skin from 3 yellow onions + 4 cups water

Bright yellow: 2 tablespoons ground turmeric + 4 cups water

Yellow: 2 tablespoons annatto seeds + 4 cups water

Teal: 1 cup fresh blueberries + 2 cups water

Magenta: 4 cups chopped beets + 2 cups water, or the juice from a can of cooked beets

Blue: Half a head red cabbage + 2 cups water

Natural Dye Uses

Painting. Pour your dyes into small bowls, add a brush, and paint with them.

Dying Eggs. Eggshells are covered in a thin protein layer called the cuticle, which has a neutral pH. Dye will not adhere to the cuticle on its own, so dip the eggs in vinegar to raise the pH and help the dye stick. Then soak the eggs in natural dyes for at least half an hour to achieve the brightest results. To create contrast, add stickers or rubber bands, or color the shells with white crayons before dipping.

Colorful Playdough. Replace the water in your playdough recipe with dyed water (see recipe on page 188).

Color Mixing. Fill small bowls with homemade dyes. Offer your child a pipette and and an ice cube tray or several bowls for color mixing experiments. Follow this up with a painting activity by giving your child a watercolor brush and heavy paper.

CABBAGE DYE MAGIC POTION

SUPPLIES
- Cabbage dye (from a red cabbage)
- Lemon juice
- Baking soda

Fill a small bowl with cabbage dye. With an eyedropper, add lemon juice to the cabbage dye and watch it magically turn pink. Now add some baking soda and watch it change to blue. You can continue to go back and forth with the lemon juice and baking soda to change the color from pink to blue.

Because blue cabbage juice contains anthocyanin, a water-soluble pigment that changes color according to its pH, it turns pink or red when it mixes with an acid such as lemon juice, and it turns back to blue (or green) when it mixes with a base such as baking soda.

Look around the house and collect other ingredients to mix with the cabbage dye. You might try soap, salt, vinegar, or clear soda water. Neutral ingredients such as water will not change the color of the juice.

For another twist on this, fill a cup with cabbage dye. Blow bubbles in the dye with a drinking straw. How does your breath affect the color of the dye?

KITCHEN CHALLENGE:
AN EDIBLE INVESTIGATION

*I have always attached great importance to the manner in which an experiment
is set up and conducted . . . the experiment should be set up to open
as many windows as possible on the unforeseen.*

—Frédéric Joliot-Curie, French physicist and Nobel laureate

Improvisation is at the heart of my point of view as a maker, and despite the many cookbook volumes I own and covet, my cooking style has always been improvisational as well. I like to stock my fridge and pantry so that I can plan meals on the fly, and I'm an enthusiastic expert at substituting one ingredient for another. This is partially due to a lack of planning, but to a great degree, it connects to a desire to learn more about how ingredients work together to make culinary magic.

Practicing improvisation in the kitchen teaches children to experiment (and possibly fail) in a low-stakes way, with the potential of eating the results of their efforts.

For this to work, it helps to make a simple family recipe with your child. Walk through and involve him in all the steps. Answer his questions, talk about how the ingredients come together, invite him to name the ingredients as you go, and ask him to think about other ingredients that might taste good in the recipe. Without letting too much time pass, follow this up with an invitation to invent a recipe using similar ingredients. This is definitely a parent-participation experience.

I've chosen pancakes as the basis for the kitchen challenge, because they're easy to mess around with without sacrificing the integrity of the recipe, but I encourage you to choose a basic dish that's near and dear to your family. Other recipes we've had success with include smoothies, pizza, salads, pastas, soups, and oatmeal.

Supplies
- Clipboard with paper and pen or sketchbook
- Measuring spoons and bowls
- Blender
- Buffet of ingredients (see the sidebar)

BUFFET OF INGREDIENTS

- Flour—all-purpose flour, wheat flour, or cake flour
- Binder—eggs, egg substitute, and applesauce
- Fat—butter, oil, and coconut oil
- Grain—wheat germ, dry oats, cornmeal, flax meal (You can replace up to half of the flour with whole grains and still preserve the integrity of this recipe.)
- Fruit—blueberries, mango, strawberries, or a banana
- Liquid—water, milk, coconut water, almond milk, soy milk, orange juice, yogurt, buttermilk, or seltzer water
- Extras—vanilla, citrus rind, chocolate chips, chocolate syrup, sprinkles, nuts, or seeds

INVITATION

Make pancakes with your child using the basic recipe provided or one of your favorites. Once he is familiar with the basic recipe, invite him to invent his own pancake recipe (or whatever else you plan to make). Talk about the ingredients and tools you'll use. Ask, "What goes into pancakes? Is there anything that you would you like to add to your pancakes that we didn't include last time? What tools will we need to make pancakes?" Help your child gather ingredients and tools (the mixing bowl, measuring cups and spoons, and so on).

Do what you can to step back and act more as a facilitator than an instructor. I often step in with gentle suggestions when it comes to quantity (like when my daughter wanted to add ¼ cup of salt), but I try to support my child's desire to make decisions as she proudly takes charge of the kitchen and learns from her mistakes. Keep the paper and pen handy for writing down the adjusted recipe so you'll be able to re-create whatever magic comes from this experience or simply preserve the recipe as a record of early kitchen inventions.

BASIC PANCAKES

This recipe is based on Mark Bittman's Everyday Pancakes, with some small adjustments.

Serves four

INGREDIENTS
- 2 cups all-purpose flour
- 4 teaspoons baking powder
- 1 teaspoon salt
- 1 tablespoon sugar (optional)
- 2 eggs, beaten
- 1½ to 2 cups milk
- 4 tablespoons oil (optional), plus cooking oil for the griddle

Whisk the dry ingredients together in a bowl. In another bowl, whisk the milk into the beaten eggs, then stir in the oil if desired. Lightly mix all the ingredients, stirring only enough to moisten the flour; too much mixing will make the pancakes tough. Add more milk for thinner pancakes.

Heat your griddle over a medium heat. Add oil to the pan. Tip: For evenly browned pancakes, soak a folded paper towel with cooking oil and evenly coat the pan with the towel.

Pour batter from a ladle or use a turkey baster to squeeze designs onto the hot griddle.

When bubbles appear on the surface of the pancakes, two to four minutes, flip them. Cook until the second side is lightly browned.

CONCOCTIONS IN A MICHELIN-STARRED KITCHEN

An Interview with Bruno Chemel, Owner Baumé Restaurant, Palo Alto, California

Bruno Chemel began his cooking career in France when he was fifteen, and he is now the master of his own creative kitchen. One of the most notable things about his menu is that it stems from an experimental approach to cooking that pulls from his knowledge of food chemistry. Chemel is a grown-up master of concoctions, and his thoughts may give us some insight into the long-term benefits of kitchen experiments and home-brewed concoctions.

Q: Your kitchen career began at a young age. Can you tell us about your early years and how you got your start as a chef?

A: When I was fifteen years old, I told my dad that I wanted to go to school to be a cook. My sister and I were independent. When I went to work, I moved away from my house, and I would see my parents on the weekends. My parents spent time with me but not too much time. No one supervised me, and it gave me a lot of clarity. The first chef I worked for influenced me a lot. He was always looking to improve himself, and it motivated me to get better. And I now see myself competing with myself. My dad always told me that the first chef I worked for influenced my future. As a kid, you mold yourself to whoever is teaching you something.

When I came home on the weekends, my mom would say, "Why don't you cook something for us?" It was almost an *Iron Chef* thing. I would use the vegetables in the fridge and send my mom to the store to pick up fish to complement the vegetables. I do the same thing in my restaurant today.

Q: How do you come up with dishes at Baumé?

A: I like to go to the farmer's market. I like the challenge of making something from one ingredient and will often begin with a starter as inspiration for a dish. I could start with a lemon and then end up making a piece of beef—with no lemon. Right now we have a lot of Sicilian pistachios in our kitchen, and I might ask, "What can we do with these?" We could make a pistachio cake or maybe a passion fruit something.

I'll start with a piece of paper and write things down. The next day, I'll look at it and say, "Ah, no." Sometimes you need other people to help you, and my sous chef will have ideas. We work as a team; I believe in teamwork. We take a lot of notes. I may get an idea for a dish, but when it's time to conceive it, I need more input than me because I'm not that good. It might take two to three days for us to put something together, and then we might not like the presentation. After a week or so, the dish is done.

We might put it on the menu for a month, or maybe we'll change it after a week if it's not working. The menu itself is composed like music. The dishes need to go together. For example, I'll start with caviar followed by egg and then move on to a vegetable and then soup. One taste inspires and complements the flavor of the next.

Q: What advice do you have for today's young alchemists?

A: I wake up every day and try to do better. I have two Michelin stars now, yes, but next year I could have zero. Every day is a new day. I'm a chef, and I try to make the best food I can every day. Life is not like a movie, it's like the theater. Every day you have to start fresh. You need to repeat and evolve at the same time. Every day we make mistakes, and we need to try to understand why we made the mistake and what we can do better.

DISCOVER

Through the process of following their curiosities, children act as scientists as they look carefully, ask questions, test, and document their discoveries. The experiences in this section take place both outdoors and indoors, as they celebrate the beauty and phenomena of the natural world.

Sensory experiences, such as splashing in a tub of water or feeling wind blow against their skin, are some of the first ways that children learn about the natural world.

When my older daughter, Nola, was one, she loved to eat blueberries and also got a thrill from smearing and splattering them all over her high chair. This activity wasn't limited to blueberries, and we saw similar Pollock-like tendencies with yogurt, blended peas, and oatmeal. The pleasure she got from squishing her meals was palpable, and I quickly spotted sensory opportunities in just about everything (including lots of nonstaining materials): ice cubes, a bowl of water, or a box full of dry beans.

The activities in this chapter will get you outdoors, taking photos, growing things, and paying close attention to details and phenomena of objects in your everyday surroundings—both natural and man-made. Through these activities, children get better at asking big questions, exploring their environment, and making comparisons.

HOW TO SET UP A
DISCOVERY AREA

An Interview with Parul Chandra, teacher,
Bing Nursery School, Stanford University

Q: The discovery area of your classroom seems to shift with the seasons, and it's always filled with inspiring objects. Can you tell us about this area and how you organize it?

A: Our discovery area consists of a low table with plenty of baskets and bins that are thoughtfully filled with seasonal and natural materials that act as provocations. We've found that the discovery area is most meaningful when the materials relate to the children's experiences, so we often display objects that connect to the time of year or include nature-based discoveries made by the children. Seasonal additions could be pinecones, leaves, shells, and different colors of sand. We include tools such as bug jars, binoculars, wooden trays with dividers, and maps for children to pick up for outdoor explorations. The trays can be numbered or categorized in different ways to encourage further classification and exploration. There are also plenty of interesting papers and notepads for children to use in recording their questions, observations, and ideas.

The table is accessible, always available, and child-centered. The goals in the space are set by the child, and we see a lot of experimentation, hands-on exploration, and understanding of materials by setting challenges and asking questions.

Scale is important. When children are young, bigger materials are easier for them to explore. Children love variations in size and scale, and it's exciting to watch them play with large pieces of cardboard or big tubes. Having a range of small, medium, and large sizes of any given object presents all kinds of interesting challenges. Similarly, we have a lot of success with materials in quantity: a barrel of apples or a large basket full of potatoes for children to sort, hold, touch, notice, and organize.

In the latter example, exploring the texture and appearance of the potatoes became a lesson when the children began to ask questions that related to their experience. After discussing the rough texture, some of the children noticed the potato eyes, and questions arose about their role. When children begin to ask their own questions in the context of curiosity, adults become coinvestigators. This is a far more powerful way to learn than if the adult just comes straight out and tells the children, "This is a potato, these are the eyes, and this is what they do."

Q: I admire how you talk with children and help them generate big ideas. Can you share some of your tips for connecting with children?

A:

- Turn off your own agenda. It's important to set aside downtime when you're not too distracted by other things that are going on in your life.
- Play with materials.
- Give yourself permission to be engaged in an activity where there's no goal.
- Put yourself in your child's shoes: jog your memory and remember your favorite childhood memories. What were they?

Q: What advice do you have for someone who would like to set up a discovery table at home?

A:

- Have a low table with bins or baskets to collect objects and provocations.
- Include tools for inquiry and exploration, such as binoculars, maps, and clipboards with pencils.
- Display, sort, and organize seasonal objects.
- Display a quantity of natural objects, such as potatoes or rocks, along with a book about each object.
- Set up an intriguing provocation for children to discover in the morning. For example, if your child shows an interest in tree branches, gather a collection of twigs, sticks, and branches. Add paper to make rubbings.

PLAYDOUGH BUILDING

Playdough is a wonderful material for improving fine motor skills, developing imagination through exploratory play, and supporting early engineering and building skills. This recipe, which happens to be the same stuff that every amazing preschool teacher I know relies on, will make enough dough for an entire preschool class. I usually make half the recipe, and it's still plenty for my two kids and their visiting friends. For more than one color, divide the uncolored dough and add color when the dough is cool to touch.

If your child is new to playdough, begin with just the dough itself, then slowly introduce tools.

Ingredients
- 5 cups water
- 2½ cups salt
- 3 tablespoons cream of tartar
- 10 tablespoons vegetable oil
- 5 cups flour
- Food coloring or liquid watercolors

- Scent, such as pumpkin pie spice (or a blend of cinnamon, nutmeg, ginger, and cardamom); ginger and cinnamon for a gingerbread scent; or peppermint, rosemary, lemon, or lavender essential oil (optional)

PREP

Mix all the ingredients together in a deep pot. To create multiple colors of dough, invite your child to help you mix everything *but* the food coloring together in a large pot until it is combined but still lumpy. Don't worry, the dough will get smoother as it cooks.

This next part is for grown-ups only. Cook the dough over low heat, mixing frequently. The water will slowly cook out of the mixture, and it will start to look sticky. Keep mixing until the edges of the dough along the sides and bottom of the pan appear dry. Pinch a piece of dough. If it's not gooey, it's ready. Remove from heat.

To make more than one color, place the dough on a countertop, large cutting board, or cooking tray that can withstand a little food coloring. Knead the warm dough until it's smooth, then divide it as many times as necessary for the number of colors you'd like to make. Roll each piece into a ball and flatten it into a disk. Add a little food coloring and knead it in. Your fingers may get a bit stained; to work around this, try to fold and knead the dough over the drops of color. Add more food coloring until you get the desired shade.

PLAYDOUGH TIP

Early Scissor Practice. Roll playdough into long worms and invite your child to cut them. Playdough is much easier to cut than paper and a great material to practice on.

INVITATION

Once your dough is ready, clear a table and place a small mountain of playdough in front of yourself and your child. It can stick to paper and is best used on a smooth surface such as plastic, wood, or vinyl. Begin by playing with the dough: poke it, roll it, stretch it, and squeeze it. Once your child has simply played with the dough, you can introduce one or two objects with which to manipulate it. Our favorites include forks, a butter knife, rolling pins, cookie cutters, a potato masher, and a garlic press.

Store the dough at room temperature in a large ziplock bag or sealed container. Unused, it will keep for months.

GLUTEN-FREE PLAYDOUGH

INGREDIENTS
- 1 cup rice flour
- 1 cup salt
- 1 cup cornstarch
- 4 teaspoons cream of tartar
- 2 cups water
- 4 tablespoons cooking oil
- Food coloring or natural dyes (see page 173)

Mix all the dry ingredients together in a pot. Add the oil and water, and mix well. Stirring constantly, cook the dough over low heat until it forms a ball, about three minutes. Cool and then knead it until it is smooth. If the dough is too wet, add more cornstarch; if it's too dry, add water. Divide the dough into the number of colors you want. Form the pieces into balls and flatten them into disks. Squeeze a few drops of food coloring onto each disk and knead it in. You can store the dough at room temperature in a sealed bag or container for weeks.

CLOUD DOUGH

Cloud dough and playdough are two of my favorite homemade sensory materials. They cost very little to make and can provide hours of entertainment. The recipe for cloud dough couldn't be simpler, as it's just flour and oil. Keep in mind that the flour from the dough can easily spread across a room, which is why we like to play with this silky substance outdoors.

Supplies
- 8 cups all-purpose flour
- Sensory box
- 1 cup vegetable oil
- A few drops lavender essential oil, available in health food stores, or other scent (optional)
- Mold-making tools and scoops such as spoons and bowls

INVITATION

Involve your child in the measuring and mixing process if you can. Pour the flour into the center of your sensory box. Create a crater in the middle of the flour and pour the vegetable oil into the hole. Mix everything together with your hands. If you wish, sprinkle in a small amount of essential oil to give it a nice scent.

Play with your cloud dough! Scoop up a handful and squeeze it in your hands. What happens? Crumble this handful up and begin again. Build a mountain and poke holes in it with your fingers. Fill a bowl with cloud dough, packing it in tight. Flip the bowl over to release the dough. What happens? After playing with the cloud dough for a bit, you might invite your child to help you find other tools for manipulating the cloud dough.

Store unused cloud dough in a large resealable plastic bag or airtight plastic container at room temperature. The dough will keep as long as the oil lasts. Give stored cloud dough a sniff to check its freshness.

POUNDING FLOWERS

This exploration will appeal to any child who likes to move and make noise. It's also the perfect spring or summer activity, when flowers are abundant, and you can teach children about the natural dyes that can be harvested from plants. We often pick our flowers from our garden, but we've also picked up wildflowers on our walks and foraged wilting store-bought bouquets on their last blooming days. After all the pounding is done, these naturally dyed beauties can be cut up and turned into gift cards or bookmarks, or simply signed and dated as a piece of stand-alone art.

Supplies
- Watercolor paper
- Collection of flowers, petals, and leaves
- Round river rock or wooden hammer
- Waxed paper
- Scissors (optional)
- Hole puncher (optional)
- Ribbon (optional)

INVITATION

Place a collection of flowers on your worktable. Invite your child to choose a flower to begin with and place it on top of a large piece of watercolor paper. Cover the flower with a piece of waxed paper and pound the petals until they make a mark on the paper. Continue pounding flowers and leaves onto the paper until you create a composition you like.

To make these into bookmarks, cut the pounded flower images into long strips, punch a hole at one end, and tie a piece of ribbon through the hole.

SCAVENGER HUNTS

Scavenger hunts are thrilling for children and help them notice details in the world around them. This is one of my favorite experiences to do with my kids. There's something about getting outdoors in the fresh air, seeing the world through their wide-open eyes, and making discoveries with them that is both fulfilling and fun. The scavenger hunts that follow help children take stock of their environment; tune their attention to the aesthetics of their surroundings; and build vocabulary that relates to photography, colors, shapes, and language.

INVITATIONS

Photo Documentary Experience. Hand your child a camera and provide guidelines on how to use it. Head off on an adventure and snap photos of anything of interest. To give this experience more direction, encourage your child to take photos of specific categories of subjects, such as red things, things that move, or objects on the ground. Consider turning your photos into a printed sheet or photo book when you're done.

Rainbow Hunt Experience. Hunt for the colors of the rainbow and take photos of what you find. Try to find the colors in the order in which they appear in a rainbow. The jury is out on the exact order, but this is my favorite: red, orange, yellow, green, blue, purple. My daughter prefers to add pink at the end. Chat about what you see along the way.

Shape Hunt Experience. Make a chart of shapes (squares, rectangles, circles, triangles, octagons), and draw or write down the names of any objects that fit the bill.

Letter Hunt Experience. This is a good one for urban walks. Make an alphabet chart and check off the letters as you spot them on signs. Or take photos of each letter.

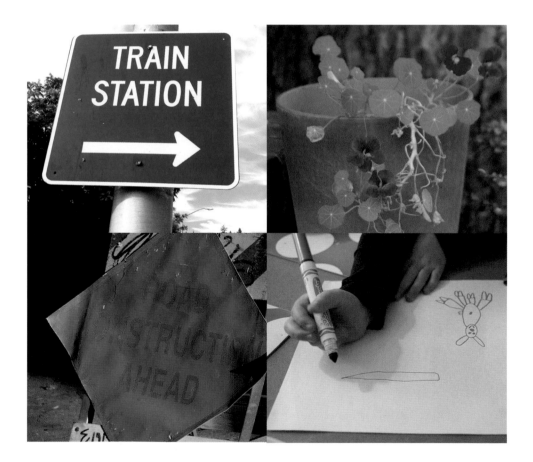

DIY LIGHT BOX

My husband and I each have memories of building our own light boxes in middle school—I built mine from a sheet of glass suspended over some thick encyclopedias and used it to trace old documents; he built one for an animation project. While ready-made light boxes are now easy enough to find in stores, large ones are pricey and take up a lot of room; besides, there's a certain thrill in putting together a light box on a dime. Deborah Stewart, a talented preschool teacher who runs a site called Teachpreschool.com, inspired me to make one for my kids from a clear storage box that we use for sensory play. Lindsey Boardman, the writer of the blog *Filth Wizardry*, inspired us to fill our light box with salt. Why salt? When you draw into a layer of salt, light shines through the thinner areas and the marks you make display beautifully. When it's not being used as a light box, we can easily disassemble it and turn it into a sensory box filled with beans and funnels.

Supplies
- Two large, clear, under-the-bed storage boxes with clear lids
- String of holiday lights, light rope, pop-on lights, or similar low-heat lights
- White tissue paper
- Paper tape, 2" in width
- Clear tape
- Extension cord

HOW TO MAKE A FLAT LIGHT TABLE

This style of light table is great for investigating dry and flat objects. To make it, remove the lid from the box and cover the inside of the lid with white tissue paper. Tape the edges of the paper to the lid with clear tape.

Fill the box with one strand of holiday lights. Dangle the end of the cord over the edge of the box and shut the lid. If the box doesn't snap shut, that's okay. Add an extension cord if needed and plug it in.

Place colorful plastic cups, sea glass, transparent tangram tiles, Magna-Tiles, or other transparent objects on top of the box.

HOW TO MAKE A DEEP-WALLED LIGHT TABLE

This style of light table is great for investigating wet materials and loose objects. You will need two boxes; remove the lids from both. Flip one of the boxes upside down. Cut a piece of white tissue paper to cover the "top" of this box. Tape the paper in place with clear or paper tape and rest the second box on top of the tissue paper, right side up. Secure the boxes together with packing tape. Run a string of holiday lights under the first box.

INVITATIONS

Ephemeral Collage. Offer your child small objects such as tumbled glass beads, pebbles, acorns, tangram tiles, colorful plastic cups, or seashells so he can make compositions with them on the flat light table. The process of manipulating and arranging small objects gives children the opportunity to explore the relationships of color, shape, and texture to create unique compositions and patterns.

Salt Experience. Pour salt into the deep-walled light box and invite your child to draw in it with her fingers. Once your child understands a little bit about how the salt and light interact, you can introduce tools such as paintbrushes and stamps. Children will quickly understand the parameters of this game and may seek out their own tools for discovery. Despite the deep walls, salt has a way of sneaking onto floors, so be prepared with a broom or vacuum.

Water Beads Experience. Fill the deep-walled light box (see photo on page 38) with hydrated water beads for a fun sensory experience (this is one of our favorite activities). Water beads are used commercially as a vase filler and can be found in the floral section of your local craft store. If the beads are already hydrated, fill your light box with them and play. If not, they will start out the size of barley; simply follow the directions on the package to hydrate the beads and then add them to the sensory tub that's on top of the light box. Watch children who tend to place small objects in their mouths closely if you're using water beads.

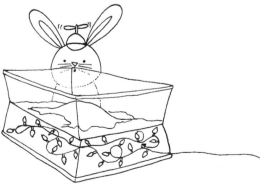

PHOTOGRAMS

These camera-free photographic prints are made on light-sensitive paper with found and foraged materials, and can help children understand negative shapes and principles of composition. I like to begin this experience with an early-morning, material-hunting nature walk. The project itself is best done on a cloud-free day when the sun is bright and high in the sky.

Supplies
- Sun print paper or fabric
- Water source
- Small natural or man-made objects
- Flat piece of clear acrylic or glass that will cover the paper or fabric

PREP

Sun print paper and fabric can be found in toy stores, art stores, or online. Keep your paper in its light-proof bag until you're ready to begin. Be sure to read the directions that come with your paper, just in case they differ from what I'm saying.

INVITATION

Find a small basket or bag, and begin by inviting your child to take a walk to collect small natural treasures such as leaves, sweet-gum-tree balls, twigs, and flowers. When you come home, find a sunny, flat, outdoor spot to set up your "photo studio."

Remove a sheet of sun print paper from the package and carefully reseal the package. Place the paper on a flat surface and cover it with a composition of your natural treasures. If you have flat items like leaves, cover them with the piece of acrylic, which will help them lie flat and keep them from blowing away.

Wait five to ten minutes, or until the sun print paper loses most of its color. Read the instructions that come with your paper for more specific details. Remove the acrylic sheet and natural objects from the paper. Rinse the paper in cool water, then ooh and aah over the results.

VARIATION

If you don't have sun print paper, you can also do this with construction paper. Place the objects on top of the construction paper (do not use no-fade construction paper), and put it in a sunny spot. The sun will bleach the color from the paper, and when you remove the objects after about an hour, you should see their shadows on the paper.

EPHEMERAL INSTALLATION

This is a fun project to undertake on a hike or day trip to the beach. Children often gravitate to making land art without any sort of prompt, but you can also direct their attention to the extended possibilities of using natural materials as well. This can be a purely aesthetic experience, but it can also be interactive when other people happen upon one of your land art creations and are challenged to see their environment with fresh eyes.

An added benefit of this exploration is that it teaches children that art does not have to be permanent and can blow or wash away with a little wind or the crash of a wave. When you make things in a public setting, take good care to respect the environment and the experiences of those around you. And be safe!

Supplies
- Natural materials
- Discovery supplies (see page 27)
- Basket or bag to collect materials

PREP

Look online or in books at the work of Andy Goldsworthy or other land artists ahead of your trip. Share and discuss the images with your child. Ask her questions like the following:

- What materials did the artist use?
- How did the artist put the materials together?
- Will the art last a long time? Why or why not?
- If you had access to these same materials, how would you arrange them?
- If you could make land art, what materials would you like to use?

INVITATION

To start, you may want to invite your child to take a walk outside to search for natural materials. Bring along a discovery pack, and use the materials in it to help you look closely at objects you encounter. The goal of this walk is to look for objects that can be turned into installation art. Some examples include rocks, leaves, twigs, and seashells.

Each unique environment poses opportunities and challenges that are different from the next. Collect your materials and choose a spot to display them. For example, create a pattern of leaves, make tall piles of rocks, or create long lines of branches that connect end to end. If other people are nearby, it can be fun to step back and enjoy watching them interact with and witness your land art.

SHADOW INVESTIGATIONS

Shadow puppetry began in China during the Han dynasty (about two thousand years ago). Shadow theater is still popular today, not just in Asia, but all over the world. My personal introduction to it was in Bali, where the puppet masters performed behind large stretched screens of fabric that were illuminated with torches. At home, the process of designing the theater and puppets is an exercise in building and digging into the imagination, and the experience of using the theater enables children to explore light phenomena while playing out creative stories.

Supplies
- Large piece of cardboard
- Scissors
- Exacto knife (for adults only!)
- White tissue paper, thin white butcher paper, or waxed paper
- Clear tape
- Construction paper, any color
- Sticks, such as popsicle sticks, straws, chopsticks, pencils, twigs, or skewers
- Floor lamp or flashlights
- Con-Tact paper or clear packing tape (optional)

PREP

Make a screen: Cut a very large piece of cardboard so it's little larger than a doorframe, then cut a rectangle from the center of the cardboard with an Exacto knife. Next, cover the opening with white tissue paper and tape the edges to the cardboard. Secure the cardboard frame to the doorframe, about a foot above the floor, with paper tape. This extra room will allow your child to move easily from one side of the screen to the other. A word to the wise: parchment paper cannot be taped. Be inventive! We have a door with a window, and we simply taped tissue paper across the window.

INVITATION

Invite your child to make puppet shapes from construction paper. To make the puppets extra sturdy, sandwich them inside two sheets of Con-Tact paper, packing tape, or a laminating sheet.

Tape a stick to the bottom of each puppet shape. Once you have enough puppets, place them behind the screen and shine a light behind them to illuminate them. If you have flashlights, prop them up on something so the puppet master can be free to work the puppets.

Perform a show!

EXPERIMENT

Play Variation. Place mystery objects behind the screen and encourage children to guess what they are.

Puppet Variation. With a black permanent marker, draw puppet images directly onto clear acetate. Tape these images to sticks.

Light Variation. Skip the screen altogether, and shine a light from behind the puppet or other precut shapes directly onto an empty wall.

Take It Outdoors: Drawing Shadows. Take some chalk outdoors on a sunny day when the sun is low in the sky—a couple of hours before or after noon. Stand on the sidewalk, or a similar concrete or tile surface, and draw your child's shadow (or vice versa). Alternatively, repeat this same exercise in the same spot at 10 A.M., noon, and 2 P.M. to document the movement of the earth.

DIY LAVA LAMP

Much like the thrill children get from mixing baking soda and vinegar, making their own fizzy, bubbling lava lamp is full of surprises and teaches them about chemical reactions.

This lava lamp requires the use of effervescent antacid tablets (I know that doesn't sound too appealing). Antacid tablets contain sodium bicarbonate (baking soda) and citric acid that react with the water to release sodium citrate and carbon dioxide gas. Water and oil don't mix, so when the antacid reacts with the colored water, it forms bubbles that rise up through both the water and the oil. When the bubble reaches the surface, it pops, and the colorful water sinks back to the bottom. Children should not eat antacid tablets, so use your best judgment and keep an eye on young children during this exploration.

Supplies
- Clear plastic bottle
- Food coloring
- Vegetable oil
- Water
- Funnel
- Effervescent antacid tablets, such as Alka-Seltzer

INVITATION

Place all of your supplies on the table and ask your child to help you fill the bottle about three-quarters full with oil; use a funnel if necessary. Fill the rest of the bottle with water, leaving an inch or two of space at the top. Wait for the water to settle at the bottom of the bottle. This might be a good time to talk about how oil and water don't mix and how oil is lighter than water. Once the water settles, squeeze a few drops of food coloring into the bottle. Do *not* mix. Cut one antacid tablet into four pieces, and ask your child to predict what will happen when you add it to the oil and water. Drop one of the small pieces into the bottle and watch it bubble. Once the show is over, you can add another piece of antacid and start all over again.

EXPERIMENT

- Place the cap on the bottle and shake it back and forth. Add an antacid tablet and shake again. What happens?
- Illuminate the bottle with a flashlight. What happens?

MYSTERY BAG

This curiosity-filled activity landed in our kitchen one rainy afternoon, and it couldn't be simpler to pull off. All you need is a deep bag and a handful of small (safe-to-touch) objects. This is a fun, playful way to explore natural and man-made items that help children problem solve through the sense of touch. My four-year-old couldn't get enough of it, and my two-year-old *had* to look inside the bag before touching everything—and that's okay too.

Supplies
- Paper or canvas bag
- Small objects, such as cotton balls, a leaf, a sponge, a ribbon, a pebble, a straw, a small pumpkin, a hard-boiled egg, an apple, an action figure, a block, and sunglasses

INVITATION

Fill the bag with about five objects. Pull everything out and invite your child to investigate the items. Put them back in the bag and ask your child to dip her hand in, feel around, describe what she feels, and guess the object. Have her pull the object out of the bag and see if her guess matches. While your child is feeling the object, ask her questions to encourage discovery, such as, "How does it feel? Is it smooth, bumpy, or rough? Does it feel like it's natural or man-made?"

EXPERIMENT

Make a Blind Discovery. After doing this a few times, your child might enjoy guessing the objects without previewing them first. Drop some new objects in the bag and try the game again.

Switch Roles. You may also enjoy switching roles—your child fills the bag with mystery objects that *you* have to guess. Playing the role of grown-up can be very empowering.

Play a Matching Game. Gather objects in twos such as two cotton balls, two apples, and two markers. Show your child all of the objects, then place them in the bag. Invite him to reach in and, using only his sense of touch, find the matching objects.

THE BENEFITS OF
BASIC MATERIALS

Jennifer Winters, director,
Bing Nursery School, Stanford University

In children's early years, the basic materials of blocks, clay, paint, sand, and water are best for fostering cognitive, social, emotional, and physical development. Resist the urge to buy all the latest gadgets and tablets that claim to promote child development. What benefits children most are hands-on, open-ended materials that offer endless opportunities for exploration. In addition, a thoughtful adult or teacher who can guide, support, and extend the child's learning will optimize the use of these ideal tools.

In the cognitive domain, children are encouraged to think, create symbolic representations, focus and concentrate, and explore scientific principles and mathematical concepts. For instance, a four-year-old announces that he wants to build the Golden Gate Bridge using blocks. First, he has to think symbolically about the elements that make up that bridge. Next, he must transfer that knowledge to the blocks to represent his idea. In doing so, challenges may arise that enable him to test principles such as balance, symmetry, serial order, and spacing. Language skills are strengthened as he is introduced to and uses new vocabulary such as *span, tension, abutment, architect, beam, concrete,* and *suspension.*

In the social/emotional domain, the skills of cooperation, taking initiative, and respecting the work of others are evident when children work with basic materials. For example, a group of children might work together to construct a castle in a sandbox or on a beach. They first decide on its basic design and general attributes, such as windows, tunnels, or even a moat. As they start to build the castle, problems or differences of opinion may arise. One child wants to embellish the castle with small berries, and another child disagrees. A compromise is worked out, and the castle is completed with berries surrounding all of the turrets.

Clearly, the benefits of self-expression, self-concept, self-confidence, and self-regulation are all strengthened when a young child has an idea; announces it to others; listens to their feedback ("That's a great idea," or "I don't like that idea; I want to make a volcano"); and responds to that feedback in a way that keeps the play going (often through compromise).

In the physical domain, young children's large muscles are first to develop. Opportunities like digging in and lifting sand are ideal for strengthening these growing muscles in the arms, legs, and torso. Balancing on their knees or standing on their tiptoes to reach something also strengthens the large muscles and increases stability and coordination. The same can be achieved easily through block building.

Fine motor skills are also increased through the use of basic materials. For instance, holding a paintbrush and moving it from side to side promotes hand-eye coordination, hand preference, and the ability to make basic strokes and shapes. Finger painting is ideal for improving hand strength, finger isolation, and thumb opposition. Manipulating clay strengthens the hand muscles that will later be necessary to hold a pencil. Water play also contributes to fine muscle development when children pour liquid from one container to another. This process requires eye-hand coordination and spatial awareness as children learn to judge when the container is full.

The opportunities for young children to develop in all the domains (cognitive, social, emotional, and physical) using basic, open-ended materials such as blocks, clay, paint, sand, and water are boundless. The cost of these items is low, and the return children receive in terms of their development is high.

ACKNOWLEDGMENTS

Endless gratitude goes to the following people:

- Scott, for all the ideas he shares with me, his devotion to our family, and his never-ending kindness. You're the best thing that ever happened to me.
- My beautiful girls, for keeping my wheelhouse running, my sink full of dishes, my floor covered in glitter, and my heart full of joy.
- My family: Sheila and Arye, for encouraging me to forge my own path and giving me tools to parent with purpose; G-Ma and TD, for providing support, friendship, encouragement, and endless baby holding—whoever said in-laws make for trouble did not know the two of you; Etan, Miriam, and Chris for inspiring me with your creativity, zest for life, and connection with my kids.
- The thoughtful and articulate educators and designers who contributed articles and interviews for this book: Jillian, Nancy, Jessica, Margie, Susan, Grace, Don, Bruno, Parul, and Jennifer.
- Elliot Eisner, for the afternoon conversations that helped me think more deeply about the field of arts education and why it matters.
- Jessica Hoffmann Davis, for her enthusiasm and the way she pushed me to think about big ideas in education.
- Steve Seidel, for introducing me to the Reggio Emilia school of early childhood education and documentation as a tool to learn about how children learn.
- MaryAnn F. Kohl, whose books introduced me to hundreds of engaging activities for my two-year-old. You are a gift!
- Jean Van't Hul, the Artful Parent, for being the creative mom I wanted to be and for making it all seem so darned easy. I blame you for everything. And I love you dearly.

- The smart and savvy Tinker-circle, who contributed their ideas to this book and gave me tremendous feedback: Melissa Allen, Angela Angaleta, Danielle Ashton, Jen Berlingo, Maya Bisineer, Cam Bowman, Marnie Craycroft, Chelsea Davidson, Victoria Dye, Jennifer Fischer, Kara Fleck, Amanda Gross, Sarah Hopkins, Cathy James, Melinda Jaz Lynde, Melissa Jordan, Rebecca Jordan-Glum, Sarah Lipoff, Paola Lloyd, Chelsey Marashian, Alissa Marquess, Allison McDonald, Rachel Miller, Amanda Morgan, Bernadette Ortiz-Grbic, Alicia Otani, Yuliya Patsay, Anna Ranson, Jamie Reimer, Jillian Riley, Cate Styer, Jill Martin Wrenn, Maggy Woodley, and Sara Yau.
- Danielle, Rachel, Aleksandra, Teceta, Tenaya, Rebecca, Jillian, Liz, Susan, Diana, Elizabeth, Aude, and Olivia, for the late-night chats and friendship.
- Shirley, Emily, and Marisa at Bumble, Los Altos, whose kindness and genuine connection with my children enabled me to write this book while saving my sanity.
- My cheerleading crew and friends at the San Jose Museum of Art, and especially Toby Fernald and Suzette Mahr, for the many rich conversations about art, family, and life.
- My publishing team: The creative crew at Shambhala and especially Jennifer Urban-Brown, the best editor a lollygagging tinkerer could ask for—this book would not exist without your vision and guidance; Erica Silverman, my agent, who handles my big questions with patience, grace, and wisdom; Chrissy Watson for taking a pass at my manuscript in the middle of parenting and writing legal briefs!
- All of the children I've had the great honor of learning from and working with.
- My incredible readers and online friends, who answered my crazy late-night questions, commented on my blog posts (even the not-so-good ones), and helped me become the best parent-educator I could be.

I'm so grateful.

THE BUSY
PARENT'S PLANNER

These are my go-to projects to occupy my children when I have a morning full of chores or an afternoon that requires a second (or third) cup of coffee to keep my eyes open. I rotate these invitations, just as you might rotate toys, and my kids can engage with them if they want to. If they would rather work on a puzzle or tinker in the sandbox, my investment is low, and I'm happy they've found something to do that interests them.

To make these invitations work, you'll want to have the materials ready ahead of time, but they're mainly things you'll have handy, so it shouldn't add too much extra work to an already trying or exhausting day.

- A sensory tub with warm water, food coloring, scoops, bowls, and funnels (see page 185)
- Playdough and dough tools (see page 188)
- A bowl of pom-poms, dry beans, and/or pebbles; a muffin tin; and a spoon (see page 100)
- A lightbox with transparent colored cups, agate slices, colorful clear tiles (like Magna-Tiles) (see page 197)
- A creative invitation with construction paper, cut pieces of some new paper (such as wrapping paper, tissue paper, phone book pages), a tape dispenser, scissors, and ribbon or string
- A creative invitation with large drawing paper, seasonal still-life objects, and mark-making tools
- Toothpicks and gumdrops (see page 102)
- Ramps and cars, marble run (see page 108)

REFERENCES

Csikszentmihalyi, Mihaly. *Flow: The Psychology of Optimal Experience.* New York: Harper Perennial, 1991.

Curtis, Deb, and Margie Carter. *Designs for Living and Learning: Transforming Early Childhood Environments.* St. Paul, Minn.: Redleaf Press, 2003.

Dweck, Carol. *Mindset: The New Psychology of Success.* New York: Ballantine Books, 2006.

Eisner, Elliot W. *The Arts and the Creation of Mind.* New Haven, Conn.: Yale University Press, 2002.

Hart, Betty, and Todd R. Risley. *Meaningful Differences in the Everyday Experience of Young Children.* Baltimore, Md.: Paul H. Brooks Publishing Co., 1995.

Hirsh-Pasek, Kathy, Roberta Michnick Golinkoff, Laura E. Berk, and Dorothy G. Singer. *A Mandate for Playful Learning in Preschool: Presenting the Evidence.* New York: Oxford University Press, 2008.

Institute for Learning Innovation. "Thinking Through Art: Isabella Stewart Gardner Museum School Partnership Program, Summary Final Research Results, 2007." Institute for Learning Innovation, Annapolis, Md. Accessed April 29, 2013. www.vtshome.org/system/resources/0000/0069/ISGM_Summary.pdf.

Kent, Corita, and Jan Steward. *Learning by Heart: Teachings to Free the Creative Spirit.* Boston: McGraw-Hill/Irwin, 2008.

Lowenfeld, Viktor, and W. Lambert Brittain. *Creative and Mental Growth.* Upper Saddle River, N.J.: Prentice Hall, 1987.

Maxwell, John C. *Failing Forward: Turning Mistakes into Stepping Stones for Success.* Nashville, Tenn.: Thomas Nelson, 2007.

Seelig, Tina. *inGenius: A Crash Course on Creativity.* New York: HarperOne, 2012.

Tharp, Twyla. *The Creative Habit: Learn It and Use It for Life.* New York: Simon & Schuster, 2005.

ABOUT THE CONTRIBUTORS

Parul Chandra was born and raised in New Delhi, India. She has a BS in history from Delhi University and an MA in early childhood education from the University of Cincinnati. Parul has been a teacher since 1986, when she started her first preschool program in a private school in New Delhi. She has been teaching at Bing Nursery School for twenty years and is a lecturer in the psychology department at Stanford University, where she teaches several courses in child development. She enjoys organizing presentations for small groups of educators and parents to highlight the value of viewing children's competencies. Parul has one son in college.

Born and raised in Moulins, France, **Bruno Chemel** developed a keen interest in cooking at a young age. He credits both his parents with fostering his culinary passion—his mother for allowing him to watch, taste, and learn at her elbow during the preparation of family meals, and his father for allowing him to enjoy the finest flavors and subtle nuances of French gastronomy in some of the best restaurants in his native country.

As a chaired senior lecturer at Harvard's Graduate School of Education, **Jessica Hoffmann Davis** (jessicahoffmanndavis.com) founded and was the first director of the arts in education program. Dr. Davis is the author of *Why Our High Schools Need the Arts* (2012), *Ordinary Gifted Children* (2010), *Why Our Schools Need the Arts* (2008), and *Framing the Arts as Education* (2005).

Elliot W. Eisner is professor emeritus of education and art at Stanford University. He was trained as a painter at the School of the Art Institute of Chicago, studied design and art education at the Institute of Design at the Illinois Institute of Technology, and received his PhD from the University of Chicago. Professor Eisner's contributions to education are many. He has been especially interested in advancing the role of the arts in American education and in using the arts as models for improving educational practice in other fields. He is the author or editor of sixteen books addressing these topics, among them *Educating Artist Vision, The Kind of Schools We Need,* and most recently *The Arts and the Creation of Mind.* Professor Eisner was president of the National Art Education Association, the International Society for Education through Art, the American Educational Research Association, and the John Dewey Society.

As the cofounder of the groundbreaking *ReadyMade* magazine, **Grace Hawthorne** has been involved with the maker movement since its early days.

Nancy Howe was born in Brooklyn and raised in a solar-heated nursery school built by her parents on Long Island. She has a BS in education from Mills College of Education in New York City and an MA in early childhood education from Lone Mountain College in San Francisco. Nancy has been a teacher for more than thirty-five years. This is her twentieth year as a head teacher at Bing Nursery School. Nancy has three grown children and six young grandchildren. She loves creating mixed-media collages with found materials collected on weekend runs to yard sales and flea markets.

Dan Klein teaches improvisation at Stanford University, where he is on the faculty of the drama department and the Graduate School of Business and teaches at the Hasso Plattner Institute of Design. In 2009 he was named Stanford Teacher of the Year by the university's student association.

Susan Harris MacKay is the director of the Portland Children's Museum Center for Learning in Oregon, which works in collaboration with Opal School to document how children create, invent, imagine, play, and learn. She is mom to three children of her own. You can find her at http://wonderlove.typepad.com and http://opalschoolblog.typepad.com.

Jillian Maxim grew up in a large family and has always felt compelled to clear clutter and keep things organized, both for herself and for others. She holds a bachelor's degree from the University of California, Santa Barbara, has travelled the world, and spent over a decade as a business owner. Currently Jillian lives in a spatially challenged home in the Bay Area with her husband and two young children. They allow her to hone her organizational skills on an hourly basis, and for that she is eternally grateful.

Margie Maynard currently lives in Sonoma, California, where she is director of education and public programs at the Sonoma Valley Museum of Art. She's been working as a museum educator since the mid-1980s and often has occasion to speak about herself in the third person. Reflecting on her years of experience, a favorite period is one she spent working at the San Jose Museum of Art with the ingenious and alarmingly productive Rachelle Doorley.

Jennifer Winters is the director of Bing Nursery School at Stanford University and serves as a lecturer in the university's department of psychology. She holds an MA in child development from San Jose State University and a BS in elementary and special education from the University of Maryland. She has been involved in the field of early childhood for twenty-eight years. She has presented at the local level (Peninsula Association for the Education of Young Children, Bing Institute, and local early childhood programs and schools), the state level (California Association for the Education of Young Children), the national level (National Association for the Education of Young Children), as well as the international level, at education forums and symposiums in Korea and China. Some of the topics in early childhood education that she is interested in and has presented on related to early childhood education are block building, dramatic play, basic and found materials, project work, the value of mixed-aged groupings, and supporting multiculturalism in play-based early childhood programs. She has served as a mentor to programs schools in the NAEYC accreditation process.

ABOUT THE AUTHOR

Rachelle Doorley is an arts educator, community builder, and founder of the popular creativity blog *Tinkerlab*. She studied costume design at the University of California, Los Angeles, and worked on Hollywood films before finding her true calling as an arts educator. After teaching art in Los Angeles schools, Rachelle earned a master's in arts education from Harvard, and then oversaw docent and education programs at the San Jose Museum of Art. Rachelle lives with her husband and her two curious daughters in the beautiful San Francisco Bay Area, where she leads workshops on visual thinking and hands-on creativity. Rachelle believes in finding fun and meaningful ways to make every day creative, and can often be found experimenting in her sketchbook, taking her kids on adventures, and asking lots of open-ended questions. Rachelle enjoys chai tea, hand-drawn letters, train travel, hikes in the woods, artist studios, and ocean air.